SCIENCE ACTIVITY BOOK

FROM THE SMITHSONIAN INSTITUTION

SCIENCE ACTIVITY BOOK

GALISON BOOKS
GMG PUBLISHING, NEW YORK

A Galison Book
Published by GMG Publishing Corp.
25 West 43rd Street
New York, NY 10036

ISBN 0-939456-51-6

Project Director: John H. Falk, Ph.D.
Authors: Megan Stine, Craig Gillespie, Laurie Greenberg,
Jamie Harms, Sharon Maves, Larry Malone,
Carol Moroz-Henry, Gladys Stanbury

Designer: Marilyn Rose
Illustrator: Simms Taback
Editor: Cheryl Solimini
Production Editor: Hillary Huber
Design Assistant: James P. Bernard
Publisher: Gerald Galison

Fourth Printing

CONTENTS

INTRODUCTION

Science and technology touches nearly every facet of our lives today. By the 21st century, our society will demand that all its citizens possess basic competencies in the fundamentals of science and the use of technology. As science increasingly becomes the dominant subject of the work place, it is important to begin developing within children an understanding and appreciation of science early in their lives.

Learning can, and does, occur in many places and many situations. Learning occurs in school, at home, and on the trip between home and school. This book provides suggestions for interactive science activities that can be done in a variety of settings, using inexpensive and readily available materials. Whether the activities are done in a classroom or in a home, they will provide adults and children with increased opportunities to explore natural phenomenon in an engaging and exciting way. Included are experiments, activities, crafts, and games that allow you, whether teacher or parent, to learn science along with your children. The only requirements for success are the directions provided with each activity, a few common household items, a little bit of time, and some curiosity and imagination. The activities in this book are designed as curricular materials, educational guides for you to use in teaching science.

SOME SUGGESTIONS FOR TEACHERS

The activities in this book should be used as supplements to your normal classroom science curricula. Since they were originally developed for use in out-of-school situations, they may require some minor modifications to permit a larger number of children to participate. Nonetheless, you will find that these activities lend themselves well to a fun-filled science lesson for all participants.

An increasing number of school districts are exploring the use of "take-home" lessons in order to build stronger learning partnership bonds between parents and teachers, home and school. These materials have proven to be an excellent source for such "team-building" efforts. Both teachers and parents find these activities rewarding ways to provide quality learning experiences for children.

SOME SUGGESTIONS FOR PARENTS

One of the most important jobs that you have, as a parent, is the education of your children. Every day is filled with opportunities for you to *actively* participate in your child's learning. Together you can explore the natural world and make connections between classroom lessons and real-life situations. You will learn the value of asking good questions, as well as strategies for finding answers to those questions.

FOR BOTH TEACHERS AND PARENTS

The best things you can bring to each activity are your experience, your interest, and, most importantly, your enthusiasm. These materials were designed to be both educational and enjoyable. They offer opportunities for discovery, creative thinking, and fun.

HOW TO USE THIS BOOK

The science activities in this book can be successfully implemented by any interested adult, regardless of his or her science background. Accordingly, the above have been designed so there is no one "correct" solution and no "right" way to do it. Do not be afraid to say "I don't know!"

There are twenty activities in this book; since every classroom and family is different, not all activities will be equally suitable. Take the time to browse through the book and find the ones that seem to make sense for your class or family. There is no prescribed order to these activities, nor any necessity to do all of them. Once you have selected an activity to do, take the time to read through it before you attempt to do it.

At the beginning of each activity is a list of all the materials you will need to do the project. Try to assemble all of these items before you begin. The procedures have been laid out in an easy-to-follow, step-by-step guide. If you follow these directions, you should have no difficulty doing the activity. Once you have completed the basic activity, there are also suggested variations that you can try, now or later. At the end of each activity is an "Afterwords" section. This section is for you, the adult. It is intended to provide additional information, not on how to help children but for the interest of an adult participant—take some time to read it for your own enjoyment.

ASKING QUESTIONS

Encourage your children to ask questions, even if you don't know the answers. The essence of science is asking questions, and then trying to find out the answers. Some of the answers can be discovered in books, some through observation, and some, at present, are unanswerable by anyone. Ask questions like:

(before you start)
"What do you think is going to happen when we do this experiment?"
(during the activity)
"What do you see?"
"Does this remind you of anything else you've ever seen?"
(after the activity is completed)
"What do you know about X now, that you didn't know before we started?"

"Is there anything you don't understand? How can we find out the answer?"

Encourage all kinds of answers, and all kinds of questions. Sometimes the crazy ones are the ones that work. Often there is more than one answer to a question, so be tolerant of diversity and open to multiple solutions. Use the library or an encyclopedia to help answer questions and further your understanding. Lead, or have a child lead, a discussion after the project is completed. This will help to pull together what happened, why things happened, and what the activity was all about. Just remember that it may take more than one exposure for some of the ideas introduced in these activities to "sink in." These activities are beginnings, not endings. Finally, don't be afraid to be a learner yourself—that was a large part of why these activities were developed in the first place. They are for learning, adults and children *together*.

John H. Falk, Ph.D.
President
Science Learning, Inc.

SUPER SLEUTH

SUPER SLEUTH

Snooping, sneaking, and sleuthing are a lot of fun, but harder than you think. It will take you 90 minutes to collect your suspects' fingerprints and solve the crime.

YOU WILL NEED

Lipstick or an inked stamp pad (Carter's Micropore or similar non-reinkable pads are easiest to clean off)
Package of 3″×5″ plain white file cards
Several sheets of 8½″×11″ paper
Pen
Paper towels, soap, and water for clean-up

There are more than 200 million fingerprint cards on file with the FBI. Do you think the police look at each one every time they want to identify a fingerprint found on a smoking gun? At that rate, the police would only solve about one crime every 10 years! In fact, there's a system that helps detectives locate fingerprints pretty quickly. Find out how fingerprints are sorted and classified by doing a little criminal-science work of your own.

FINGERPRINTING THE SUSPECTS

1 Make and label a bunch of blank fingerprint cards like the ones shown on this page. You'll need two blank cards for each person in your family—one for the right hand and one for the left hand.

2 For the next few steps in this activity, you'll want to have a sink, paper towels, and soap nearby. Starting with your right hand, ink your thumb on the stamp pad, or lightly cover your thumb with a thin coating of lipstick. Be sure to get the ink or lipstick on the *sides* of your thumb—not just on the flat part. To do this on the ink pad, you'll have to roll your thumb from one side to the other. You can help one another with the finger rolling.

3 Now practice making a thumbprint by rolling your thumb the same way on a piece of white paper. If you can clearly see the details of your thumbprint on the paper, you're ready to begin filling in your fingerprint card. If not, keep practicing until you can make good, clear fingerprints without smearing.

4 Now ink your thumb again and roll it on the square labeled "thumb" on the Right Hand card marked with your name. Continue making fingerprints on the cards until you have a complete set—right and left hands—for each person in your family.

HINTS FOR SUCCESS

■ You should ink only one finger at a time—otherwise you'll smear ink all over the cards by accident.

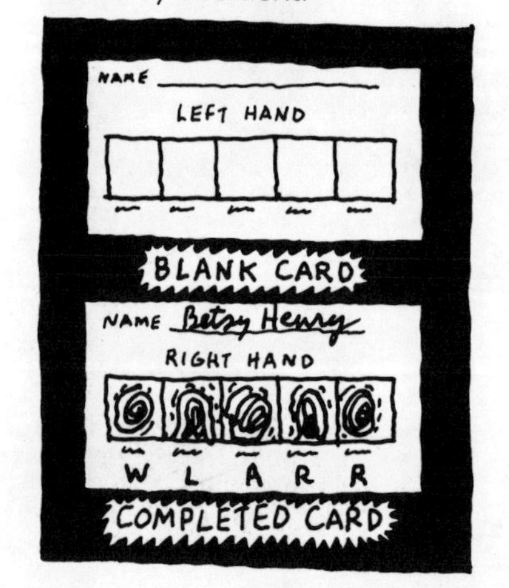

- When you roll your thumb or finger from one edge to the other in the ink, press *lightly* so that you don't pick up too much ink.
- Press *firmly* when you roll your thumb or finger onto the labeled card. That way, you're less likely to move your finger and smear the print.
- If the "ridges" on your fingers don't show up well when you ink them, soak your fingers in hot water for a minute or two before trying to ink them again.

CLASSIFYING THE PRINTS

5 Look at your fingerprints and compare them with the print types shown in the chart. Decide whether each print is an Arch, a Whorl, a Right Loop, or a Left Loop. Sometimes a fingerprint will look like an Arch to one person and a Whorl to another, so discuss the classifications with everyone in your family and go with the majority vote. If you can't agree on a type for a particular fingerprint, mark it with a U for Unknown.

Mark the first letter of the print type under each fingerprint on your card. Use A for Arch, W for Whorl, and so on.

COMMITTING THE CRIME

6 Choose one person to be the detective and send that person out of the room. Decide what kind of crime you want to commit, and choose someone to be the criminal. You can make up a story to go along with the crime, but the important thing is that the criminal must leave his or her fingerprints on a piece of white paper. Draw a picture of a weapon, a drinking glass, an envelope, or whatever else might go along with your crime scenario, on a piece of white paper. If you can't think of a crime, you might draw a picture of a safe, like the one shown here.

7 One by one, ink the fingers of the criminal's right hand, and put his or her fingerprints on the drawing of the weapon, safe, or whatever. Be fair about it, or your detective will never solve the case! Put the criminal's fingerprints on the drawing in the right order—thumb, index finger, middle finger, ring finger, and then pinkie. Secretly send the criminal to wash his or her hands. (Or better yet, have *everyone* go to the sink, just so the detective won't overhear anything and guess who the criminal is!)

8 Now call the detective back into the room and let him/her classify the prints on the drawing, writing the appropriate letter under each print. When all five prints have been classified, the detective should be able to look at the sequence of letters and match it to an *identical* sequence on the fingerprint cards you've all made.

VARIATION

Real fingerprints are invisible — until you dust them with a fine powder to make them show up. Find out how much harder real-life detective work is by collecting and classifying some real-life prints!

- Scrape some pencil lead across an emery board, letting the fine black powder fall into a dish.
- Use a soft watercolor brush and very carefully "dust" the powder onto a drinking glass you've touched. The powder will stick to the oil mark your fingers made.
- Place a piece of clear tape on the dusted print, press lightly, and then lift the tape off. Save your lifted fingerprint by taping it on a clean card.
- Play Super Sleuth the *hard* way. When you're out of the room, have a member of your family secretly leave a single "mystery" print on a clean glass. See how long it takes you to find its match on the fingerprint cards.

AFTERWORDS

Long before fingerprints were accepted as a means of identifying criminals, a Frenchman named Alphonse Bertillon developed an identification system of his own. Bertillon's method, adopted in 1882, was based on the idea that each person's physical measurements were unique. When a criminal was brought to the French police division where Bertillon worked, he would take a set of eleven measurements, including the length of the torso, hands, legs, feet, etc. These measurements remained on file and helped the police identify repeat offenders.

Although Bertillon's system was widely used for about 30 years, it was eventually abandoned because it wasn't accurate in every case. Fingerprints then became the preferred method of identification. William Herschel is credited with developing a practical fingerprinting technique, which he used in Bengal, India, in the 1850s, even before the Bertillon system was adopted in Europe. But Sir Edward Richard Henry, who was Herschel's successor in India, came up with a way to *classify* the prints.

The classification system developed by Sir Henry is still in use in America today. It consists of a series of numbers and letters written like a fraction, with the symbols for the right hand appearing above the line and the symbols for the left hand appearing below. For instance, a typical fingerprint code would look like this:

$$\frac{5\ \text{Ar-r}\ 12}{17\ \text{T-2r}}$$

The first number in the classification is a code that tells whether or not there are whorl patterns in the set of prints. Other letters stand for various pattern types, and some parts of the classification even involve counting the ridges in a particular print! This is a slow, painstaking process, best done with a magnifying glass.

When the police find a fingerprint at the scene of a crime, the first thing they do is classify the print. If there is a suspect handy, the police can fingerprint that person and carefully compare the prints to see if they match. But more often than not, the suspects are *not* handy. Then the police must turn to the fingerprints that are *already on file.* Even these filed fingerprints would be useless if they weren't classified. Sir Henry's classification system is so complete that police can quickly find a similar set of prints by simply looking through the code numbers. Then, and only then, the careful visual identification process begins.

Of course, fingerprints aren't the only unique physical characteristics that can be used to identify people. Newborn babies are always footprinted immediately after birth, to ensure that there will be no "mix-ups" in the nursery. The noseprints of dogs are equally unique, and so are certain boney projections on horses' legs. But our Super Sleuths are content with a system that's right at their fingertips.

SUN TRAP

SUN TRAP

A Sun Trap takes from 60 to 90 minutes to build. It also takes a sunny spot to succeed. A bright yard, fire escape, or window sill will do.

YOU WILL NEED

1 Cardboard box
about 14″ × 12″ × 10″
8 Empty soup-size cans
4 Clear plastic bags
or plastic wrap
2 Toothpicks
Sharp knife or scissors,
masking tape, newspaper,
ruler, pencil, water,
aluminum foil
Room thermometer,
rubber bands (optional)

At last you can do something about the weather. You can make it warmer, cooler, wetter, or drier in your own solar greenhouse or sun trap.

A solar greenhouse collects the sun's heat during the day and stores it until the heat is needed at night. Because it can store heat, a sun trap doesn't need much extra fuel to keep warm.

You can build your own solar greenhouse with these plans. In your own sun trap you can grow tropical plants, heat water, and dry fresh fruit.

CUTTING COMMENTS

It takes a lot of cutting to make a sun trap. You can use a scissors, but the best cutting tool is a sharp knife. When cutting with a knife, always try to:

- cut away from yourself
- place your fingers behind the blade
- keep the knife in a safe spot.

Never use a sharp knife without the permission or supervision of a responsible person.

1 If the cardboard box is not square, choose one of the longer sides to be the front. Label the sides with a pencil.

- Cut the top off the box.
- On the left front side, draw a line from the top left corner to the bottom right corner. Cut along this line.
- On the right side, draw a line from the top right corner to the bottom left corner. Cut along this line.
- Cut along the bottom front edge. Remove the cut pieces from the box.

2 Cut the bottom edge of the left side to the back. Cut the bottom edge of the right side to the back. Tilt the back wall of the box forward about two inches. This will direct more sunlight to the center of your greenhouse where your plants will be.

3 Carefully, turn the box over. Mark the excess cardboard on the right side. Cut off the excess. Mark the excess cardboard on the left side. Cut off the excess.

4 Tape the bottom edges together on both sides with masking tape. Be sure the edges are airtight. Cut off the extra cardboard across the bottom.

5 Cut vents in the middle of the bottom of the left and right sides. Open the vents so that the flaps are on the outside. Vents will help cool your greenhouse in the summer and give plants the fresh air they need.

6 Tape several layers of newspaper to the outside of the back wall for insulation. Insulation will help keep your greenhouse warm in winter and cool in summer.

7 Cover the inside of the back wall with aluminum foil. This reflector will help aim sunlight toward the center of your greenhouse where the plants will be. It will also help warm the heat storage cans.

8 Tape together several clear plastic bags or pieces of plastic wrap to cover the front of the box. Tape one side to the bottom of the box. This plastic will let sunlight in and help keep heat in your sun trap. It will also help keep cold air and wind out, just like the glass in a greenhouse.

9 Fill the cans with water. Cover the top of each can with plastic held down with tape or a rubber band. These heat storage cans will collect and store heat from sunlight during the day. They will slowly give off that heat to warm the greenhouse at night.

10 Take your box, cans and tape outside to a sunny, dry, level spot. Even the edge of a front step will do. Face your greenhouse toward the sun.

- Line with water-filled cans along the back wall of your sun trap.
- Place the thermometer, if you have one, on the floor of the greenhouse to check temperature changes.

- Tape the plastic over the front. Stretch it as tightly and smoothly as you can. This will let the most light through.
- Close the vents and be sure your greenhouse is airtight.

If your greenhouse gets hotter than you want it to, prop the side vents open with toothpicks and unfold the top of the plastic cover one inch. Please note that your sun trap is not waterproof. Do not leave it outside in the rain or snow.

You can use your solar greenhouse for more than growing plants outside. Use it for starting seedlings, growing an herb garden, experimenting with heat, helping bread dough rise, and even drying fruit.

MAKE YOUR OWN RAISINS

You can turn your sun trap into a fruit dryer and make your own dried fruit. First, empty your sun trap.

- Cut out the left and right sides leaving a one-half inch border. Cover the two side openings with cheesecloth.
- Use cake cooling racks or other screen racks to make shelves inside. Lay whole seedless grapes or apple slices on the shelves.
- Close the front plastic flap. Leave your sun trap in the sun all day, but bring it inside at night. In a few days, the fruit will be dried and ready to eat. Store fruit in an airtight jar until needed.

VARIATIONS

- Set a can on a large piece of paper in the sun. Trace the can's shadow every hour. Decide which way your sun trap should face to catch the most sunshine.
- Find four empty cans of equal size. Place an empty can, one filled with water, one with soil and one with rocks in the sun. Cover each with plastic and a rubber band. When they are warm, bring them indoors. Which stays warm the longest?

AFTERWORDS

As early as the first century A.D., Romans were growing fruit and vegetables in simple greenhouses. Large tubs were covered with clear sheets and heated with decomposing manure. A more elaborate greenhouse built at the same time was covered with rough glass and fitted with a hot air system. It was unearthed at Pompeii.

In Victorian England, greenhouses or conservatories were added to the homes of the wealthy. They were large enough to hold fruit trees. Parties were held in these rooms so guests could pick their own dessert. Later, some families had up to a dozen greenhouses: one for melons, one for strawberries, one for exotic giant water lilies, and so on.

Today, greenhouses are still used for growing tropical plants in non-tropical climates. They are also used for germinating seeds, extending the growing season, and other agricultural needs. Agriculturists grow many types of produce in greenhouses that they are unable to grow outdoors. Agriculturists have also been able to grow larger quantities of produce. It is possible to grow thirty times more lettuce per acre in a greenhouse than outside in a field. Fruits and vegetables grown in a greenhouse need much less water than those grown outdoors. Outside, one ton of tomatoes needs about 162,000 gallons of water. The same amount grown in a greenhouse would need only 11,700 gallons. That's a savings of 150,300 gallons of water! This savings alone would be reason enough to put all farming under glass—except for one thing. Many greenhouses have little insulation and heat storage. To grow one ton of tomatoes, you would need as much as 100 times more oil to heat a greenhouse than would be needed outdoors.

In the 1930s and 1940s, greenhouse builders in New England sank their buildings partially in the ground to save fuel. They knew the surrounding earth would insulate the greenhouses.

Other fuel-saving methods were found. Gardeners learned that brick walls, rock beds, and barrels of water all act as "heat sinks," storing heat to keep a greenhouse warm after a sunny day. This was the beginning of "solar" greenhouses.

Solar greenhouses have been designed to collect and store energy in many ways. With rising heating and food costs, solar greenhouses have been put into use as simple home extensions. Many homeowners have turned the south-facing wall of their homes into efficient and attractive sun traps.

SAY CHEESE

SAY CHEESE

In only an hour and a half, you can make fresh cheese, but you must start the project a day in advance. For best results, read through the recipe before you begin.

YOU WILL NEED

½ Gallon very fresh whole milk
¼ Cup cultured buttermilk
Salt
Large Pyrex or other heat-resistant bowl
Large pot (big enough to hold the large bowl)
Cheesecloth
Colander
Knife, spoon
Plastic wrap
Small bowl

If you accidentally leave a carton of milk out on the kitchen counter all day, what happens to it? Is it spoiled? Sour? Unsafe to drink? Bacteria in the milk are multiplying and dividing as fast as they can. But don't think all bacteria are harmful. The bacteria that make milk sour are *not* harmful. Actually, they are essential to making yogurt, sour cream, and many kinds of cheese.

Here's a recipe for making soft, fresh cheese from milk that's been left standing too long. The first time you try it you might get a very soft cheese like cream cheese. Or you might get firmer chunks, or "curds," which means that you've made cottage cheese. Either way, you'll be surprised how easy it is to make a delicious and nutritious food.

BEFORE YOU BEGIN

To go on with Steps 2, 3, and 4, you'll need to be around at just the right moment, when the milk has "clabbered." The amount of time this takes will depend on two things: how much buttermilk you use and the temperature of the "warm spot" where the milk is standing (have someone check this with a room thermometer). The chart at right will help you plan ahead by showing *approximately* how many hours it will be until the milk clabbers and the curds are ready to be cut. (But check the milk early.)

Temperature	Clabbering Time
60°F	44 hours
70°F	23 hours
80°F	17 hours
90°F	13 hours

1 Let ½ gallon of fresh milk stand in its carton for several hours, until it reaches room temperature. Pour it into a clean, large Pyrex bowl and set the bowl in a warm place. Add ¼ cup buttermilk to the bowl and stir well. Cover the bowl with plastic wrap. Don't bump, move, or jiggle the bowl after this. Let it stand undisturbed overnight.

2 Within the next 24 hours, the milk will become soft, like a custard. This is called "clabbering." When the milk has clabbered, you will see the watery-looking whey collecting on top of the curd and at the sides of the bowl. Now it's time to cut the curd into cubes. Use a knife to make cuts, about ½ inch apart, through and across the curd, to form squares. As you cut through the curd, the "custard" should split and separate into distinct ½-inch-square cubes with clean, sharp edges. If that doesn't happen, your clabbered milk is not quite ready to be cut. Wait a little while and then test it again.

3 After you've cut the curd, let it sit for 20 minutes. Meanwhile, fill a very large pot half full with hot tap water and set it on the stove. Then gently carry the bowl of curds and whey to the stove and set it inside the big pot of water. Now you must heat the curds *very slowly* to make them coagulate (clot) more and to release more of the whey. Use a low gas flame or low setting on an electric stove and bring the temperature of the curds and whey up to about 100°F. If you don't have a thermometer, ask an adult to stick a clean finger in the whey. At 100°F it will feel quite warm, but not hot enough to make the person jerk the finger out. Be sure to heat the curds *slowly:* It should take *at least 30 minutes* to bring them up to 100°F.

4 Remove the bowl from the big pot. Pour the hot water out so you can use this pot to drain the cheese. Set a colander inside the big pot, and line the colander with two layers of cheesecloth. When the curds and whey have cooled slightly, about five minutes, pour them—gently—into the colander to drain. Occasionally pick up the four corners of the cheesecloth and shake the cheese, so that more whey can drain off. If you are making cottage cheese, you can twist the top of the cheesecloth "bag" and squeeze.

5 Remove the cheese from the cheesecloth and put it into a small bowl. Add salt to taste — between ½ and 1 teaspoon should be right. Refrigerate your cheese immediately, and then enjoy it!

Caution: Although the bacteria in the milk are harmless, other types of bacteria may get into your cheese before you are finished making it. *If at any time your cheese looks spoiled or smells bad, don't eat it.*

HINTS FOR SUCCESS

■ If your milk doesn't clabber within 24 hours, it probably never will (unless you have left it standing at 60 °F; see the Clabbering Times chart). You may have to start over if that happens. Use more buttermilk next time, and/or place the bowl of milk in a warmer spot.

■ Don't break the curds or stir them while you are cooking the cheese. The smaller the curds are, the more whey they lose. If they lose too much whey, your cheese will be too dry.

AFTERWORDS

There are more than 400 kinds of cheese made throughout the world. But, believe it or not, all cheeses — from Swiss cheese to blue cheese to Camembert — start out as milk and undergo the same basic procedures you experimented with in Say Cheese. Some cheeses are cooked a little longer than others. Some curds are stirred and some are not. Some are pressed only lightly, and others are put under 100 pounds of pressure for several days. And some cheeses are eaten right away, while others are left to cure for two months or more in deep, cool cellars or caves. All of these changes in method have an enormous effect on the final taste and texture of the cheese. But perhaps the most important factor is the ripening or curing process; that's when the bacteria can *really* go to work. Bacteria are the heart and soul of cheese. Without them, milk wouldn't get past the liquid stage.

The bacteria present in milk, fresh from the cow, are called *lactobacilli*. When they feed on *lactose* (milk sugar), they don't get fatter — they just divide in half and multiply. During their feast, lactobacilli give off a waste product called *lactic acid*, which first sours milk and then causes it to coagulate. (Yogurt and buttermilk are made this way.) Later, when the coagulated milk has been made into cheese, the bacteria continue to break down the protein in the cheese. This improves the texture and taste of the cheese.

However, most of the bacteria present in milk are destroyed during the pasteurization process. But 1% of them do remain, and those can be encouraged to multiply by adding some similar bacteria from a "starter" culture. When you made cheese, you added buttermilk as a starter culture, and you really didn't know what kinds of bacteria you would get. Cheese manufacturers, however, want to choose their bacteria very carefully. They have many different species, each one suited to making a particular kind of cheese.

As cheeses ripen and cure, they are often allowed to form their own rind, which seals the cheese and prevents other, unwanted bacteria from penetrating it. But sometimes — as in the case of blue cheeses such as Roquefort, Stilton, and Gorgonzola — mold is either encouraged or even added to the cheese on purpose. For instance, moldy bread crumbs are mixed into the curds to make Roquefort. On the other hand, Gorgonzola develops its blue-green veins all by itself when the cheese is pierced through to allow air to enter.

Of course, when it comes to cheese, the age-old question is: "How do they put the holes in Swiss cheese?" For the answer, just look for cheese's best friend, bacteria, again. Certain types, or strains, of bacteria produce gas, which gets trapped in little pockets in the cheese. As the gas expands, the cheese is pushed out of the way and holes form. In fact, you can tell how old a piece of Swiss cheese is by looking at the size of the holes! If the holes are small, as in Baby Swiss cheese, you'll know that it was cured for a shorter period of time; it will also have a milder flavor. For a stronger flavor and aroma, choose a Swiss cheese with larger holes.

BUBBLE BREW

BUBBLE BREW

Who doesn't like bubbles? Bubbles can be very scientific, but best of all they are fun to make and watch, which should take about 45 minutes.

YOU WILL NEED

Some thick liquid dishwashing detergent (Joy works well)
Newspaper
Measuring cup
Flat plate or tray
Soda straws
2 or 3 Empty tin cans of the same size
Can opener
Masking tape
Styrofoam or paper cups (1 per person)
Wire coat hanger
Paper clips, ruler, tap water
Heavy cotton cord or string, 3 feet long
Karo syrup, Jell-O powder, Certo, salad oil, or food coloring (optional)

1 To concoct your Bubble Brew, put six ounces of water in a cup and add one ounce of dishwashing liquid. Any time you want to make up some Bubble Brew, use six measures of water to one measure of detergent. Add the detergent to the water and stir gently.

2 Get ready to bubble! Bubbles are fun, but they can get messy. If you are working indoors, take some newspaper and cover the floor and tables so they won't get wet. You can work outside if there isn't too much breeze.

3 Try blowing half-bubbles. Smear some Bubble Brew on a flat plate or tray (or the tabletop, if your parents don't mind it getting wet). Dip one end of a soda straw in the Brew and hold it just above the wet table or plate. Blow gently into your straw. How large a bubble can you make? What happens when you touch a bubble with a wet straw? A dry one?

4 Try blowing another bubble *inside* your first one. Keep your straw wet. How many bubbles can you blow inside each other? Make a small bubble right beside a bigger one. Let them touch. Which bubble bulges into the other? Is it always the same-size bubble? Blow several bubbles and look for a place where four walls meet. Build some bubble houses and towns. **If all this blowing makes you feel dizzy, take a break!**

5 **Big Bubbles!**
Use a can opener to cut the ends out of two or three tin cans, then tape them together. Tape some plastic soda straws across the joints to make your tin-can tube stronger. If it's okay with your folks, wet the whole tabletop with Brew. Dip one end of your tin-can tube in the Brew. Make sure the Brew forms a window over the end when you take it out. Blow gently through the cans, without touching your lips to the cut metal. How big a bubble can you blow on the table? Get a friend to help you. Look at your images in the bubble.

USE SAME SIZE CANS-OPEN AT ALL ENDS
TAPE OVER STRAWS & JOINTS

6 **Floaters!**
Stand away from the table. Blow as big a bubble as you can out of the end of your tin-can tube. Gently raise your tube and twist it at the same time. Practice until you can cut off the bubble. Try making bubbles by blowing through a pencil hole in the bottom of a Styrofoam or paper cup, after dipping the open end in Brew.

7 **Super-Big Bubbles (Lucky Loops)**
Make up enough Bubble Brew to almost fill a large tray. Run about three feet of heavy cotton cord or string through two milkshake straws. Knot the cord and move the knot inside one of the straws. Put your Lucky Loop into the tray of Brew. Get it all wet—your fingers, too! Hold onto the straws and slowly lift your Lucky Loop out of the Brew.

2 PAPER CLIPS
STRAWS

Practice moving your loop through the air to make *super-big* bubbles! Walk quickly and see how long a bubble you can make. Twist the two straws toward one another to cut off a bubble. Practice! Practice! Practice!

Try the same thing with a loop made from a coat hanger. Hold one finger at the bottom of the hook. Remember, keep everything wet with Brew, including your finger.

8 Other Bubble Makers

You can make bubbles with anything that has a hole right through it. Try using a funnel, a piece of hose or plastic pipe, a piece of window screen, a cardboard roll from wax paper or foil, or whatever else you can find.

9 Bubbles That Aren't Round?

Cut some soda straws into pieces about 3 inches long. Clip two paper clips together and stick each free end into a separate piece of straw. Use more straws and clips to build structures like these. Get

some wire you can twist into different shapes.

What shape of bubble can you make with these? What does your image look like in these "windows"? Dip each one into the Brew. Get it all wet. Let windows form in all spaces. What happens if windows happen to touch? Break one window at a time with a dry straw.

VARIATIONS

- Experiment with making stronger bubbles. Try adding different amounts of Karo syrup, or Certo, or Jell-O powder to your Bubble Brew. Try adding some salad oil to your detergent before you mix it in the water.
- Can you draw a map of something you cannot see? Release some bubbles outside and watch where they go. Now you can draw a map of air currents you cannot see.
- What colors can you see in your bubbles? Where do they come from? What can you do to change the color of light hitting the bubbles? The color of the solution? Try using colored cellophane and food coloring to find out.
- How long can you keep one bubble? What can you do to protect it? Set up a contest with rules. Do bubbles shrink in the refrigerator?
- How high can you make a half-bubble? How wide? Use a wet ruler to find out.

AFTERWORDS

Liquids are made up of small particles called *molecules*. You can't see them, not even with a microscope. There is a layer of molecules at the surface of a liquid, hooked together just like links in a chain. Now think of when you pull on a string or a chain. You are putting *tension* on it. The chain of molecules in a liquid is also under a kind of tension, called *surface tension*.

If you want to see the effect of surface tension, try this: Dip one end of a dry soda straw into plain water. Hold it upright and see how much water is left in the straw. It doesn't all run out! Next add a few drops of detergent to the water and try again. How much of this solution runs out the bottom of the straw? You'll find that the detergent lowers the surface tension of the water and the solution flows more easily than plain water.

Adding detergent to water also makes it easier to blow bubbles. See the difference it makes if you use rain water to make Bubble Brew instead of tap water. Rain water has lower surface tension than most tap water.

You may have noticed that it was difficult to make a square bubble by blowing through a tube. Even if you used a square wire frame, you won't get square bubbles. Liquids tend to form drops in the shape of balls, or *spheres*. The *energy* in the surface layer of molecules is trying to make things easy for itself. The best shape for holding the most liquid within the smallest surface is a sphere. What shapes do water droplets form on the waxed surface of a car?

Did you notice that whenever four bubble walls come together, one of them usually breaks? That's because three walls coming together are more stable, or steady, than four. For instance, you may have seen lots of four-legged tables that wobbled. But have you ever seen a wobbly three-legged table? Probably not.

Whenever you make a bubble next to another bubble of a different size, the wall of the smaller bubble always bulges into the space of the larger bubble. In other words, the smaller bubble always wins the shoving match. You see, the *pressure* inside smaller bubbles is greater than the pressure inside larger bubbles. Remember how hard it is to blow up a balloon at first? But the bigger the balloon gets, the easier it is to keep blowing it up. Pressure is the *force* (push or pull) on a surface. The same force (your breath) is spread over a larger surface area in a bigger bubble than in a smaller one. That is why the smaller bubble has greater pressure.

Did you see all of the colors of the rainbow in your bubbles? These colors are contained in sunlight. You saw the *visible spectrum* of white (seen) light: red, orange, yellow, green, blue, violet. You usually can't see the separate colors in sunlight. You need water droplets (rain) or a thin film (soap bubble) to sort them out. The business of sorting out the separate colors of the spectrum is called *dispersion*. Oil on a puddle will also disperse the rainbow colors in the visible spectrum. Where else have you seen these colors? Maybe you'll find a pot of gold at the end of *your* rainbow!

MAGNETIC PERSONALITIES

MAGNETIC PERSONALITIES

Do you want to go north, south, east, or west? This one-hour activity can head you in the right direction—by teaching you how compasses work.

YOU WILL NEED

20 or 30 Little magnets (see **Note**)
5 or 6 Plastic soda straws, jumbo size
Thread
Paper clip
Cardboard box
Needle
Scissors
Tape (Duct tape works best.)
Steel sheet of some kind (baking sheet, cookie-tin lid) or iron frying pan
Small plastic container with lid (like a ½-pound margarine tub) or a clear plastic soda bottle
Felt-tip marker

Note: Radio Shack sells three kinds of little magnets: doughnut-shape ones, rectangular ones with holes in the center, and tiny ones with no holes. Get 10 of each if you can, but be sure to get several of the smallest.

Since their early discovery, the mysterious properties of magnets have served people in many ways. Magicians have used them to dazzle audiences with seemingly impossible stunts. Magnets are also necessary for navigating ships and planes: They are what make compasses work. Share in some of the mystery and make a compass of your own.

1 Put the doughnut magnets on a pencil. Reverse every other one so that all of them *repel* (push away from) each other. Hold the pencil upright, like a flag pole. Now hold the pencil horizontally and compare the spacing between the magnets. How do you explain the difference?

2 Amaze your friends! Carve away all but one layer of the cardboard in a small area on one size of a cardboard box *from the inside.* Tape a doughnut magnet there. Tape a few more doughnuts to it. Turn the box over so the magnets in the box are a secret that only you know about. Tie a paper clip to a piece of thread, and tape the free end of the thread to the table. Stick the clip to the box, right where the magnets are hidden, and carefully push the box away until the clip is suspended in thin air!

3 Magnets will either *attract* or *repel* each other, depending on which "poles" are brought together. Can you figure out whether it is two *like* poles or two *opposite* poles that attract? **Hint:** You'll need *three* magnets to solve this little problem.

4 Tape 3 or 4 magnets, one each, to the end of 3 or 4 straws. (Or just slip doughnut magnets over straws and wrap tape around the straw ends so the magnets won't fall off.) Use equal lengths of thread to suspend the straws in space from the end of a ruler. Make sure they

are all positioned exactly in a line so that the magnets *repel* when they hang next to each other. If you suspend them above a steel sheet or iron pan, you can set other magnets underneath; these will also affect the swinging magnets. Experiment to see how long you can keep the magnets swinging. If you place them just right, you may discover *perpetual motion*—the magnets may swing forever!

5 Now make your own compass. Cut a 2″ or 3″ piece of straw and cut a point on one end. Cut a notch in the center of the piece as long as the thickness of 2 little round magnets.

6 Poke (or drill) a tiny hole in the center of the bottom of the plastic container and a little hole in the center of the lid, too. Thread the needle with about 18″ of thread, and run it up through the hole in the bottom of the container. Tape over the end of the thread *and* the hole securely.

7 Trap the thread between two of your little magnets, run the needle through the middle of the notch in your straw pointer, and out through the hole in the lid.

8 Push the pointer down over the magnets, and push the whole thing down into the center of the container. Fill it with water, put the lid on, and tape the thread on the top. Don't pull

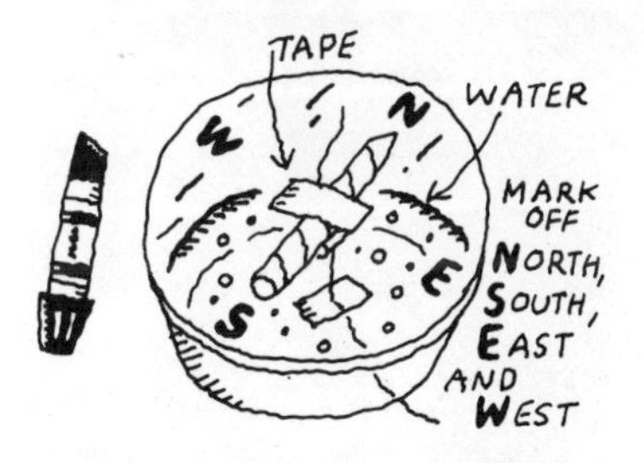

the thread very taut! (It's a little more difficult, but you can follow the same procedures to build your compass in a plastic soda bottle.) **Note:** The compass *will* work without water in the container, but water keeps the pointer from swinging too much.

9 Now you've finished your compass! Because the pointer always points north, you can figure out the other directions and use a felt-tip marker to mark off north, south, east, and west on the compass lid. Then organize a treasure hunt! Give clues with directions like "Go north to tree. From there, go west to door." Let a friend borrow your compass to find the treasure.

AFTERWORDS

The Greek island of *Magnesia* gave magnets their name. The legend goes that in ancient times shepherds grazed their sheep on this island. Each shepherd carried a wooden walking stick, or *staff,* and covered the end with iron so it wouldn't wear out so quickly on the rocky ground. The shepherds then noticed that some of the tiny stones in the soil would cling to the iron tips but would not stick to the wood itself.

Later in history but still a long time ago, Norsemen (people from what is now Norway, Denmark, and Sweden) invented the earliest compasses. The Norsemen were excellent sailors who knew how to guide their ships by the stars—especially the Norse, or "North," Star. But if the stars were hidden by fog or clouds, this *stellar navigation* was impossible. Then the Norsemen noticed that a certain type of rock, if allowed to move freely, would always point north, toward the North Star. One of the earliest compasses was made from a piece of this *lodestone,* or "leading stone." The rock was set in a block of wood and then set afloat in a basin of water. As the earth's magnetic field pulled on the lodestone, the wood would turn to point north—the direction of the magnetic field.

Other *magnetic substances* act like the lodestone. These include *ferromagnetic* metals such as iron, nickel, and cobalt. *Ferro* means "iron," and as it turned out, lodestone is a highly concentrated form of magnetic iron ore. By combining some of the simpler ferromagnetic metals, modern man can create powerful magnetic materials.

If you pick up a magnet shaped like a candy bar, you will find that its magnetic effect is mainly found near the ends—the *poles* of the magnet. There are two types of magnetic poles, and one of the laws of magnetism states that two *like,* or similar, poles will repel each other and two *unlike* poles will attract.

Finding out which pole is which can be a little tricky. First you must know that the half of the Earth nearest the North Star is called the Northern Hemisphere. The Earth's axis "sticks out" in the Northern Hemisphere at the *geographic North Pole.* One pole of a free-moving magnet will always point toward a spot nearby this North Pole. This spot is called the *magnetic North Pole.* The end of the magnet that points toward this pole is called the *north-seeking pole.* But, because unlike poles attract, the north-seeking pole of the magnet must actually be the *magnetic South Pole.* Whew! The best way to avoid confusion is to remember that all magnets in the Northern Hemisphere are labelled "north."

Even scientists do not yet completely understand magnetism. They believe that magnetism is determined by the tiny particles, or atoms, that make up the magnetic substances; pairs of electrons orbiting around the nucleus of the atom may cause magnetism. This may be why, if you take your candy-bar magnet and break it in half, you end up with *two* magnets, each with two poles.

Magnetic substances that keep this magnetism going in one direction are called *permanent magnets.* The Earth itself is a huge permanent magnet; its *magnetic field* acts between its magnetic North and South Poles. A compass works because its magnetic needle is attracted by the Earth's magnetic field.

Permanent magnets are a part of your everyday life. They are at work in electric motors, stereo speakers, telephones, televisions, and even in some refrigerator doors!

PIGMENT PIZAZZ

PIGMENT PIZAZZ

Pigment Pizazz takes about an hour to complete. It is most successful when Boston fern or Wandering Jew plant trimmings are used.

YOU WILL NEED

Handful of plant trimmings
1-to 2-Quart stainless steel or enamel pot with lid
Clear drinking glass
Stainless steel or plastic spoon
Scissors, stove, strainer, water
As many of these "chemicals" as you have in your kitchen: baking powder, baking soda, citric acid, clear soft drink, cream of tartar, orange juice, onion juice, pickle juice, salt, sugar, and vinegar

There is more color to a plant than meets the eye—more than the greens and browns of its leaves, stems, and roots. Hidden beneath a rich, green disguise, plant pigments of many colors live inside plants. Pigments are the colored substances that help plants absorb light. You can release some of them by gently simmering a few leaves in water. Add a pinch or two of kitchen chemicals and watch those pigments pizazz.

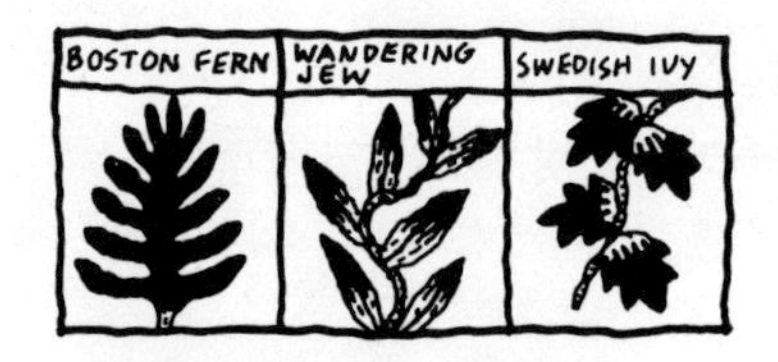

BEFORE YOU START: Picking a Plant

Choose a plant that needs trimming. If you have no plants, get trimmings from the plant of a friend or relative. If you have never trimmed a plant before, get help from someone who has. Trimming should encourage plant growth, not kill the plant. Get trimmings from these plants:

Wandering Jew (*Zebrina pendula* and other varieties)

Boston fern (*Nephrolepsis exaltata*)

Coleus (*Coleus spp.*)

Swedish ivy (*Plectranthus australis*)

Ivy and hairy-leaved plants will not work as well as the other plants. Many other types of plants will work but may be poisonous.

Note: Many plant pigments will stain clothing and wooden or aluminum kitchen tools. Wear an apron to protect your clothing and wash all kitchen tools immediately after using them.

1 Look closely at your plant trimmings. What colors do you see?

- Use the scissors to cut the trimmings into ½-inch pieces. Put the pieces into the pot.
- Cover the trimmings with one inch of water. Put the lid on the pot.
- Gently simmer the trimmings on the stove for about 20 minutes. What color do you think will simmer out of your plant trimmings?

2 While the trimmings are simmering, test your kitchen chemicals. Testing unknown chemicals can be dangerous. These kitchen chemicals are all known foods, so you can smell and taste them without fear.

- Smell the chemicals and choose the best-smelling one.
- Taste one of the chemicals. Does it taste sour or bitter? Rinse your mouth with water. Taste another. Pick out the most sour and the most bitter.

3 After 20 minutes, look at your trimmings. Is the color what you expected?

- Carefully strain the plant juice into a glass. Do not taste this plant juice—it may be poisonous. Also, clean the pot well before using it again.
- Look at the plant juice. Does your plant have more than just green pigments in its leaves and stems? The color you see comes from other pigments in the leaves and stems.

4 Pour a little plant juice into a couple of cups. Add the best-smelling chemical to the juice in one cup. What happens to the juice?

- Add the sourest chemical to another cup of juice. What happens? Compare the sweet and sour mixes.

■ Add baking powder to another cup of juice. What happens?

■ Add baking soda to your sourest chemical cup. What is the result?

■ Experiment with the other kitchen chemicals. Can you get the juice to bubble? Make your own colorful combinations.

VARIATION

Make a plant dye by following Pigment Pizazz steps 1 and 3. Just a few trimmings will make enough plant dye to color a handkerchief. After simmering the trimmings, let the juice, or dye, cool.

Drop a handkerchief or wool yarn into the dye. First, tie the fabric into knots to get a tie-dyed effect, if you choose. Stir it around to coat the whole piece evenly. Let it sit in the dye overnight.

In the morning, carefully take your cloth out of the dye and hang it out of direct sunlight to dry.

AFTERWORDS

Green leaves, yellow daffodils, and red tomatoes wouldn't be possible without plant pigments. Neither would many important plant functions. Here is why: Every color absorbs different wavelengths of light. Light energy triggers many different plant functions. For example, chlorophyll, the green pigment in plants, absorbs the light energy that triggers photosynthesis. Photosynthesis is the process by which plants make their own food. Without green pigment, there would be no photosynthesis. Without photosynthesis, plants would starve to death.

Other colors, or pigments, absorb the light energy that makes plants flower. Still others attract bees to spread the pollen necessary to produce new plants. To study the makeup and uses of pigments, scientists perform experiments like Pigment Pizazz. As you have seen, taking the pigments from plants can produce surprising results. The red flower heads of an amaryllis plant produce yellow plant juice, while the red leaves of a weeping cherry tree produce a soft green color. Different parts of the same plant can produce different colors, too. The roots of a cherry tree produce a reddish purple, but its bark produces tans, red, and oranges. Tree bark, onion skins, marigold flowers, cranberries, walnut hulls, and sassafras roots all make dazzling colors and dyes.

Not only can one plant produce many pigments, but many colors can be produced from a single pigment. Metal compounds, acids, and bases can all change the color of plant pigments.

Metal compounds such as alum can brighten or completely change the color of a plant pigment extract. They also help dyes "take" to yarns and fabrics, and prevent fading.

Acids, such as vinegar (acetic acid), will also change the color of plant juice. The white vinegar sold in grocery stores will turn some plant pigments redder. If you don't like the redder color, it can be reversed by adding a base, the opposite of an acid. When the right amount of a base, such as baking soda, is added, the plant juice will return to its original color. Adding even more base may turn the juice greener or bluer.

As you may have observed while performing Pigment Pizazz, acids taste sour and bases taste bitter. But the human tongue is not the most accurate or safe way to test for acids and bases. A more precise and safer way is to use an acid/base or "pH" indicator. Most paper and liquid "pH" indicators originate from plants. Litmus paper, used by chemists, is made from certain plants called lichens. Your kitchen acid/base indicators of vinegar and baking soda are not as sensitive as litmus paper. But even your own Wandering Jew or Boston fern can help you tell that orange juice and black coffee are acidic, while baking soda and milk of magnesia are basic.

HOME MOVIES

HOME MOVIES

Lights, camera, action! You can be a big-time movie-maker by creating your own cartoons and movie "projector," in about 30 minutes.

YOU WILL NEED

1 Empty plastic 1-gallon jug (the kind that bleach or detergent comes in)
2 Tiny screw eyes
Ruler
Dark or black pen (Felt-tip type is best.)
Sharp pocket knife or utility knife
Several rubber bands
Cellophane tape
White paper
Thumbtack
Plain white 3″ × 5″ index cards

When the lights go down at the Saturday movie matinee and the curtains slide open, an animated cartoon usually starts off the show. Or maybe your favorite cartoons are on TV on Saturday morning. Now you can have fun watching cartoons any day of the week, whenever you want. You can make your private movie studio and "screening room" at home—and learn how your eyes let you see pictures "move."

1 First soak or scrape any labels off the jug. Then draw 2 lines all the way around the jug—one about 3½″ from the bottom and the other about 5½″ from the bottom. Use a ruler, or try this: Stack up some books until they measure about 3½″ high; then lay your pen on top of the books and rotate the bottle carefully against the pen point. Add more books and repeat the process for the 5½″ line.

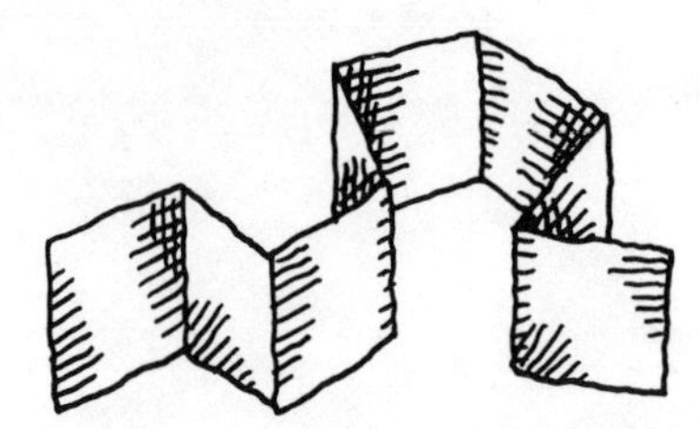

2 Cut a strip of paper several inches wide and exactly long enough to reach around the jug. Fold the strip in half, and in half again, and once again. When it is unfolded, you should see 8 equal segments.

3 Wrap the strip around the jug again (maybe tape it there) and make a mark on the jug at each paper fold, making sure that the marks are between the two lines at 3½″ and 5½″.

4 Use the knife to carefully cut a slot from line to line, ¼″ wide, at each of the marks you just made on the jug. Then cut the top *most of the way off,* just below the level of the handle. *Leave about 2″ uncut right below the handle.*

5 Screw one of the screw eyes into the center of the jug cap and the other screw eye into the center of the bottom. Your cartoon "projector" is now completed.

6 Next, cut several 3″ strips of paper just exactly long enough to form a loop that will fit *inside* the jug. (You might have to tape a few strips together to get the right length.) For most

jugs, the size will be 18" by 3". Fold the strips into eighths, just as you did in Step 2. Copy one of the sample movie strips onto one of your paper strips. Form the strip into a circle, making sure that the drawings face the *inside,* and tape the loop ends together. Drop the loop into the jug and push it down to the bottom. Reattach the top of the jug to the bottom with strips of tape.

7 Using rubber bands tied together and attached to the screw eyes, suspend the jug between two points, above and below. For instance, you can use a thumbtack to hang the jug from an open doorway, then put a chair underneath and attach the bottom of the jug to the chair's backrest. Then stand on the chair, turn the jug around a few times, and release it. Look into the jug through the rotating slots and watch your homemade cartoons move! That's *animation.*

8 Now that you have the equipment and the idea, make up some strips of your own. Everyone in the family can make up a cartoon or two. Here are some hints for success:

- Start with a simple movement, like a ball bouncing up and down. Try to "see" the motion in your mind before you draw.
- Be sure to break the movement down into several small steps so that the final motion will be smooth.
- If parts of the picture should *not* move (like the outline of the face of a winking person), it's best to draw it once with a dark pen, and then trace that part on each section of the movie strip, or "frame." Now that part will appear to be motionless in the cartoon.

COPY THESE SAMPLE MOVIES ON A STRIP 18" X 3".

VARIATIONS

- Make up a bunch of blank paper strips and invite neighbors and friends in for an afternoon of cartoon making and viewing. Show some of your favorites to start off the show, and then teach the guests how to make their own.
- Make a larger movie projector out of a 5-gallon bucket, a plastic wastebasket, or 5-gallon commercial ice-cream tub. With this larger container, you can cut *twice* as many slots and double the number of movements that you can put into your cartoon. What's the advantage? Longer features! And with more movements, your cartoons will have more realistic motions. Experiment with this new "feature-length" film studio.
- There is another kind of homemade animated feature—the flip book. It works on the same basic optical and mechanical principles you've just learned. To make a flip book, you will need a pack or two of plain white 3″ × 5″ index cards. Staple about 30 to 35 together along one 3″ edge. Draw a series of illustrations on the pages, and flip through the book at a medium speed. Now try a faster flip, and then a slower one. Which works better: the flip-book cartoons or the jug-theater cartoons?

AFTERWORDS

The movie machine you just made is a simple version of a device invented in the 19th century called a *kinetoscope*. The principles involved are very similar to those that apply to the modern-day movies that we all enjoy.

If you were to put one of your eight-frame cartoon strips inside of a transparent bottle and rotate it, all you would see is a blur. The key is to separate each image in some way so that your eye actually sees a series of single images in rapid succession. If the images flash quickly enough, one after another, our brain does not register the "dark" spaces between the images. In your bleach-jug kinetoscope, the "dark" spaces that separate the images are provided by the plastic sections between the slots you peek into.

The motion pictures that you see at the theater or at home rely on two inventions: the movie camera and the movie projector. The movie camera records individual, still images on a long strip of film, just like a snapshot camera does, except that a roll of movie film is much longer. The main difference is that a movie camera doesn't take just one shot and stop. It takes one picture after another as long as the shutter trigger is held down. Usually the camera records 24 pictures each second!

The other device that brings the motion picture to life is the projector. The projector uses a bright light to illuminate the images. Then a mechanism advances the film in front of the light and a shutter interrupts the beam of light after each individual image—24 times each second. A lens focuses the images on a wall or special screen so you can view the movie.

The last phenomenon that makes the movies possible takes place in the human brain. The retina of the eye can retain an image briefly after it has gone from the screen. This "afterimage" dissolves into the next image that flashes on the screen, giving the illusion of smooth motion.

Specialized movie cameras can play time tricks on us, either speeding time up so that we can see things happen that normally take a long time, or slowing time down so that we can observe events that usually happen too rapidly to see. Would you like to see all the leaves on a maple tree turn bright autumn colors, and then fall off of the tree—all in a few minutes? Simply train your movie camera on the tree and expose one frame of film every 5 minutes during the daylight hours in the month of October. When you show the film at the usual 24 frames per second, you can see the whole fall drama in a matter of about 3 minutes!

For the slow-motion effect, you need a camera that can expose film at the rate of 240 frames per second. Then you can film the flight of a hummingbird for three seconds, capturing a total of 720 images. Projected at the standard rate of 24 frames per second, all those fancy wing beats that happen in three seconds can be slowed to fill 30 seconds. This is something that can't be seen with the human eye.

You won't be able to do all of these special effects at home, but your simple bleach-bottle kinetoscope is a good start. Next stop: Hollywood!

CRYSTAL GARDEN

CRYSTAL GARDEN

Grow a colorful and crazy garden of crystals—in a bowl of rocks! It will take 20 minutes to set up your garden, and one or two days for the crystals to grow. While they're growing, you can do two more experiments to discover the secret nature of crystals.

YOU WILL NEED

1 Cup Epsom salts
½ Cup water
Food coloring
2 Small, shallow bowls about 5″ in diameter
A few small rocks
Saucepan, spoon, stove
Magnifying glass (optional)

If you were on a scavenger hunt and had to find a crystal, what would you bring back? Rubies? Diamonds? A chandelier? Actually, there are plenty of crystals right in your own kitchen—for example, in your salt shaker and sugar bowl.

Salt and sugar crystals are so small, you might not have noticed how beautiful they are. By doing these easy experiments, you can watch some ordinary ingredients grow into jewels!

1 Place a few small stones in each of the two shallow bowls. In a saucepan, heat ½ cup water. Slowly add 1 cup of Epsom salts, stirring constantly. Make sure the salts dissolve completely, but do not let the mixture boil. Add a few drops of your favorite food coloring and stir again.

2 Now you can experiment with this mixture to find out how temperature affects the growth of crystals. Pour half of the mixture over the rocks in one bowl, and pour the other half into the second bowl. Set one dish in a warm place—over a heater or radiator, for instance. Put the other dish in a cool, but not cold, location. *After this, don't touch the bowls.*

3 Look at your Crystal Garden 6 hours later. Have the crystals started to form yet? Soon you will see a thin crust forming across the top of the mixture. Let the Crystal Garden sit *undisturbed* overnight.

4 The next day, gently break the crust on the bowl that was placed in a cool spot. Pour off any remaining liquid. What kind of crystals do you see? Are they large or small? Square or needle-like? You may want to use a magnifying glass to observe the crystals more closely. Wait another day, and then repeat this step with the bowl that was placed near the heater. How are these crystals different from the ones that were allowed to cool and form more quickly?

VARIATIONS

■ Find out whether crystals will grow in a bowl of marbles or nails. Will the crystals grow if you don't put anything else in the bowl with the Epsom salts mixture?

■ Try growing crystals from alum or sodium bicarbonate. Both should be available at your pharmacy.

ROCK CANDY RECIPE

You can grow big rock candy crystals and eat them, if you're willing to wait a week or more.

Caution: This activity needs close adult supervision. Remember: Sugar syrup gets very hot. Handle it carefully!

■ First, suspend a weighted string inside a glass, as shown in the illustration. You can use a steel nail, a button, or several paper clips as weights, but don't use anything made of lead.

■ Next, stir 2½ cups of sugar into 1 cup of water in a saucepan. Don't be surprised if all the sugar doesn't dissolve until after you heat it up—that's the idea. Heating the water

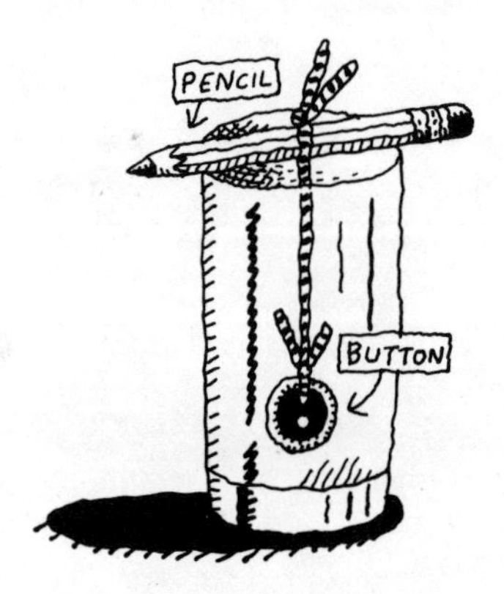

causes the water molecules to move farther apart, and this makes room for more sugar. This is called a supersaturated solution. (In fact, the Epsom salts solution you made was a supersaturated one, too.) Heated, the water holds more sugar than it does when it's cold.

■ Cook the sugar syrup until it boils. Continue cooking over medium heat *without stirring* for three or four minutes.

■ Remove the syrup from the heat and let it cool for a minute or two. Then pour the syrup into the glass with the weighted string. Be careful, because if the syrup is too hot, the glass might break. Using a towel or potholder, place the glass somewhere where you and your family can see it easily. Let it stand *undisturbed* for a week or more, until the crystals have formed around the string. Be patient and *do not move the glass.*

If the whole glass of syrup crystalizes into one solid lump, you probably cooked the syrup too long or stirred it while it was boiling. Try again!

You'll enjoy watching the crystals every day, as they grow from tiny, perfect "starter" crystals to larger, more irregular chunks. How does the sugar crystal shape compare with the Epsom salts crystals you grew last week?

■ When the crystals are large enough, break your rock candy

into pieces and enjoy it—a little at a time! Let's face it—a lot of sugar just isn't good for your teeth.

LIGHTS! CAMERA! FAKE IT!

Did you ever wonder why stuntmen never flinch when they jump through a window or get hit over the head with a bottle? The answer is fake glass—and since it's made with plain old sugar, it's one of Hollywood's cheapest tricks.

To make fake glass, grease a cookie sheet and put it in the refrigerator to chill. Put ½ cup of sugar in a saucepan and heat it over a *very low flame* until the sugar melts and turns golden brown. (Sorry—homemade fake glass isn't clear.) **Be very careful and do not touch the syrup.** Sugar gets very hot! When the sugar melts, pour it onto the cold cookie sheet, tilting it to spread the syrup thin. Let cool. This is sugar glass—just like they use in Hollywood! It might be fun to smash your sugar glass into the kitchen sink and pretend you're in a smash hit Hollywood movie!

What's the difference between sugar glass and the rock candy you made? Why didn't the sugar glass form into crystals? If you think about your experiment with Epsom salts, can you guess the answer?

AFTERWORDS

Isn't it strange that the chandeliers and goblets we ordinarily call "crystal" are actually only glass? True crystals are very unlike glass, both in the way they are formed and the way they look. By definition, a crystal is any substance in which the atoms come together in a regular, organized pattern. The atoms in salt, for example, will always arrange themselves in exactly the same way, forming a six-sided box. The atoms in sugar form a different pattern, creating a rectangular crystal that is slanted at both ends. Each and every kind of crystal, from a ruby to a grain of sand, has its own unique arrangement of atoms that result in a variety of crystal shapes. There are short, fat crystals and long, spikey crystals. Others are flat and sheet-like, or small but intricate like a snowflake. However, one thing holds true for all crystal types: They all have flat sides, which are called faces. And regardless of the crystal's shape, the angle between each set of faces is always the same.

For crystals to form, the atoms in the crystal substance must be able to move around freely, so that they can arrange themselves in their particular pattern. This is possible only when the crystal substance is dissolved in solution, or heated to a liquid, or "molten," state. As you saw when you grew Epsom salt crystals, the time it takes for the solution to cool affects the size of the resulting crystals. When a solution is cooled quickly, the atoms arrange themselves into many smaller crystals. When a solution is cooled slowly, the atoms arrange themselves into larger crystals. Once a solution has completely cooled, the atoms are no longer free to move around and the crystal-forming process stops. When the earth's crust cooled millions of years ago, this same principle applied. The crystalline materials near the surface cooled more quickly than the matter deeper underground. Consequently, the crystals that formed near the surface were rather small, and the ones underground became large rocks.

But what happens when a crystal substance is melted and then cooled so quickly that the atoms don't have time to arrange themselves at all? What happens is shown by your sugar-glass experiment; in fact, that's also what happens when sand is made into glass. Glass is a very thick, super-cooled substance in which the atoms are disorganized—just like the atoms in a liquid are. Unlike the atoms in crystals, the atoms in liquids form no particular pattern and are constantly moving around. For this reason, strange as it may sound, glass is sometimes called a liquid by scientists because it has all the properties of a liquid. And even though the atoms in glass are moving around *very* slowly, they are still moving enough to allow the glass to *flow* like a liquid, over a period of time. If you've ever seen a 200-year-old house with its original windowpanes intact, like the ones at Colonial Williamsburg in Virginia, you can see that the glass is actually thicker at the bottom of the pane than at the top. Scientists think this is because the glass is very slowly flowing downward. But it would take an *eternity* for a windowpane to flow into a puddle of glass!

SOUND MACHINE

SOUND MACHINE

The hour that it takes to make your stringed instrument will provide many hours of fun for you and friends as you experiment with sound.

YOU WILL NEED

1 Piece of smooth wood (used for shelves) ¾″×10″×16″
10 Short (1″) nails
2 Yards of clear nylon string or fishing line (strong enough to support about 20 pounds; this is available in most hardware stores)
1 1″ wooden block
Hammer, scissors, pencil
Ruler, empty 12-ounce soda bottle

Have you ever been hit by a sound wave? If so, you probably didn't mind it. No, this is not one of those corny jokes like seeing a cigar box or butter fly. A sound wave is a kind of movement that travels through the air or other substance and is picked up by our ears as sound. These movements can be small waves that move very fast and come very close together. They can move slowly and be far apart. You will be able to hear the difference between them by building a stringed instrument. Then just pluck the strings and let the sound waves hit your ears.

1 On your piece of wood, make 5 pencil marks in a straight line, beginning about 1 inch down from the 10 inch side, and about 1 inch apart. Cut the nylon string into 5 pieces of the following lengths: 14 inches, 12 inches, 10 inches, 8 inches, 6 inches.

2 Make a slip knot on each end of each piece of nylon and, leaving the knots loose, put a nail into one loose slip knot on the longest piece of string and pull it as tight as you can. Then at the first pencil mark at either edge of the wood, hammer the nail into the wood (about ½ inch). One half of the nail will be showing.

3 Put a nail into the loose slip knot on the other end of the thread and pull it as tight as you can. Then pull the string straight and very tight. Hammer the nail into the wood at that point (about ½ inch deep).

4 Repeat Steps 1 through 4 with each piece of nylon. Use a shorter piece each time, until you have used the shortest available string, which is 6″.

HOW TO MAKE DIFFERENT SOUND WAVES

When each string is tightly in place, your instrument is ready to play! Pluck each string separately with your second or third finger. Does each one make a different sound? Which string makes the lowest sound or has the lowest pitch? Which one makes the highest sound or has the highest pitch? You will notice that the longer the string, the lower the sound or pitch. The shorter the string, the higher the pitch. The longer strings vibrate, or move back and forth, more slowly than the shorter ones. The waves or air movements made by the longer strings are wider or farther apart. The waves or air movements made by

the shorter strings are faster and closer together.

■ Put the little 1-inch block of wood underneath the middle of the longest string. Pluck the string on both sides of the block. Has the pitch changed? Move the block along underneath the string. See what happens when you pluck the string on both sides after each move. Move the block back and forth and play a tune. You probably can if you use all the strings. Each time you put the little block of wood under a string, you make two shorter strings, and the sound waves for each one move more quickly. The pitch you hear is higher each time.

VARIATIONS

■ Hold a 12-inch or 18-inch ruler firmly on the edge of a desk so that about 10 inches of the ruler sticks out over the edge of the desk. Now pull the free end down gently and

let it go. Repeat the pull, but do it harder. The more energy you use, the greater air disturbance or more sound waves there are. The more energy, the louder the sound. Now try pulling the ruler down with less of the ruler sticking out over the edge of the desk. Does the sound change? Is it higher or lower? You can probably understand why it is higher.

■ Get an empty soda bottle (12-ounce, without a cap). Hold the bottle up to your mouth. Press your lower lip gently against the neck of the bottle and blow sharply across the open top of the bottle. Do this several times until you are able to make a good "toot" each time. Then put about 1 inch of water in the bottle and blow again. Has the sound changed? Add a little more water and blow again. Each time, the sound is higher.

■ Ask 4 or 5 friends each to get an empty bottle and join you. One bottle should stay empty. Put a different amount of water in each of the other bottles. Now get your Bottle Band together and play some tunes. Just by changing the amount of water in each bottle, you can change the pitch of each bottle, and blow a different note.

AFTERWORDS

Have you ever seen an eardrum? Probably not, but each time you plucked a string on your stringed instrument or blew across the top of a soda bottle, you disturbed the air and made waves that finally hit your eardrum. Your eardrum then began to vibrate or to move back and forth.

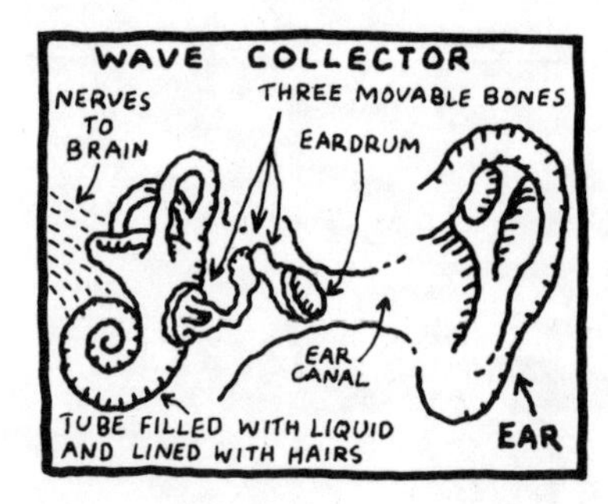

When an object, say a pin or a rhinoceros, moves, it bumps into the small particles or molecules that surround it. These molecules may be air, liquid, or solid. Before these moving molecules can bounce back, they bounce into other molecules, which then bounce into more molecules. While each molecule only moves slightly, the sound of a pin dropping or a rhino snorting may travel a long distance over a road of moving molecules. Finally, the molecules reach your ear. They bump into your eardrum, which begins to vibrate. Your nerve endings carry these vibrations to your brain and your brain translates them into sound.

Sound waves are a series of movements in molecules. If you could see sound waves, they would look like the rings of ripples made when a stone is dropped into water. Sound ripples spread from the source of movement or vibration—the dropping pin, the snorting rhino. When molecules bump together, they cause compressions; when they move apart, they cause expansions. You can picture this when you pluck your stringed instrument. As the nylon string is stretched up, it moves the air molecules above it (compression) and thins out those below it (expansion). As the string snaps downward, it does the opposite: It compresses the air molecules below and those above it expand. As they expand, they compress the molecules beyond them. As long as the string vibrates, you can hear the air disturbance as a "twang."

If we could see sound waves, they would look like ocean waves. The top or crest of the wave would show compression. The bottom or trough between would show expansion. The taller the wave, the louder the sound. The faster the waves, the higher the pitch or frequency. Your ear is shaped like a funnel to collect sound waves. It funnels the waves to your eardrum. The waves set the eardrum in motion and it pushes against three small bones. The last bone is attached to a coiled tube filled with liquid. As the eardrum moves, it sets the bones moving. The last bone sets the liquid in motion. The liquid pushes against tiny hairs that line the tube or ear canal. These hairs are connected to the nerves that transmit the rhythm, speed, and strength of the movement to the brain. Then we say that we "hear."

How the brain translates this movement into what we hear is not yet understood. It may be that what we know as sound is something we invent inside our heads.

EYE CATCHERS

EYE CATCHERS

Your eyes help you to know what is going on around you. But sometimes your eyes can play tricks on you. Investigate your eye, and its tricks. This activity will take about 40 minutes.

YOU WILL NEED

Ruler
2 Cereal boxes, paring knife
Scotch tape, aluminum foil, wax paper
Large safety pin or small nail
Magnifying glass, if you have one
7 Pennies
Penlight
2 Pencils, paper, cardboard, pushpin
Green, black, and orange paints, crayons, or felt-tip markers
Flashlight or table lamp

1 Is seeing believing? Take a good look at the top hat at right. How does its width compare with its height? Measure it and see. Surprised?

2 It's a topsy-turvy world! Things aren't always as they seem. Cut the flaps off one end of a cereal box and tape a sheet of wax paper over the opening. Make a hole about the size of a dime at the center of the other end of the box; tape a 1″ × 1″ piece of aluminum foil over the hole.

Use a big pin or small nail to poke a hole through the foil covering the hole. Light a birthday candle. (Be careful to keep hair and clothing out of the way.) Put the pinhole near

the flame on the wax-paper screen. Move the box back and forth to sharpen the image. Blow gently toward the flame. Which end of the flame is up? Your iris works the same way with light rays coming from a distance.

Now, hold a magnifying glass at arm's length and look at something 2 or 3 feet away. Which way is up? The lens of your eye does the same trick when you look at something nearby.

3 Do you have smart pupils? Your pupil looks like a little black circle in the center of each eye. It

controls the amount of light that enters. Stand in front of a friend who is looking straight ahead. Turn on your penlight and bring it slowly from behind your friend's head around to the front. What happens to the size of your friend's pupil as the light gets closer? As it moves away?

4 Light that enters your eye through the pupil is focused by the lens to form an image on your retina. But there is a spot on the retina, your blind spot, or *optic disk,* where nerves and blood vessels enter. Light that is focused on the blind spot cannot be seen. Find your own blind spot by making a black pencil mark the size of a dime on your paper. Make a small star about 3″ to the right.

Hold your paper about a foot away and stare at the star with your left eye and shut the right. Move the paper slowly toward you, and then back. When does the dime disappear? Or you can put 7 coins in a row on a table. Put your face 6″ to 8″ away and stare at the center coin. Close your left eye. Which coins seem to disappear? Now, try the other eye.

5 You can get an idea of what your blind spot and the tiny blood vessels in your retina "look" like. Shut one eye and touch the tip of a lighted penlight to the upper eyelid. Keep the tip moving gently in a tiny circle. What sensation do you get? Which part of your eyelid gives the best results? What happens when you stop moving the light?

6 Your retina has *photoreceptor* (light receiving) cells, which respond to incoming light. *Cones* are cells that are especially sensitive to bright light and colors. Cones are more numerous near the center of the retina. *Rod* photoreceptor cells are scattered all over the retina, but especially *away* from the center. Rods are very sensitive to dim light and help you to see at night. Here's one way you can test how rod cells work.

Stand behind a friend who is looking straight ahead. Slowly and quietly move a pen or any small object from about 2 feet behind the head in a circle around to the front. Ask your friend to say "Stop!" as soon as he or she first can tell anything is there. Gradually bring the object forward. When can your friend see black and white? Colors? Identify the object? How good is your *peripheral* (side) vision?

7 Hold a pen or pencil in each hand and at arm's length. Relax and bend your elbows just a bit. Close one eye and try to touch the two pencil points together. Now try it with both eyes open. Which way works best? How good is your *depth perception?* Next, put one end of a soda straw on the tip of your nose and point it away from you. Try to focus on the far end of the straw. How many straws do you see?

8 Images don't disappear right away after you're finished looking at something. Sometimes, one image is still causing a sensation (hanging in) and another image is placed on top of it. You

may have seen a Western movie where the wheels on a wagon seemed to be going backward. Here's how you can get that effect.

Color a 4″ circle of white cardboard with black, as shown above. Fasten your pinwheel onto the end of a pencil eraser with a thumbtack. Turn on your TV and spin your pinwheel in front of the screen as you watch the spokes. What do you see?

9 The black-and-white pinwheel gave you black and white *afterimages.* You can also have afterimages in color. Try this. Draw an American flag on a 4″ × 6″ piece of white paper or cardboard. Color the flag so there are 7 *green* stripes alternating with 6 *black* stripes and 50 *black* stars on an *orange* rectangle. Stare at your flag for 30 seconds or more while shining some extra light on it. Then quickly stare at a blank, light-colored wall or big piece of white paper. What happens? How long does the sensation last?

AFTERWORDS

Who makes a living by causing our eyes to play tricks on us? Magicians do! They use nimble fingers and distractions to lead us into seeing what isn't there. To help you understand how this "magic" works, try a trick with your friends.

First, be sure to wear a shirt or sweater and have a 25¢ coin ready. Then bend your left elbow and rest your fingertips on your right shoulder, near your collar. (Your elbow should be pointing down.) Tell the audience that you'll make the quarter disappear by rubbing it into your elbow.

Put the three middle fingers of your right hand together and lay the coin across the tips of them. Warn your audience that they should keep their eyes *on the quarter.* Then rub it in an up-and-down motion on the surface of your elbow. *Deliberately* let the quarter slip and fall to the floor. Pick it up in your *right* fingers and begin rubbing your elbow once again. Let the quarter drop at least two more times. By then your friends may call you clumsy—but agree with them.

Now suppose you have dropped the coin for the third time. This last time, quickly retrieve the coin with your *left* hand (even though all along you have been picking it up with your right). But keep the three fingers of your right hand close together as always.

Once you pick the coin up in your left hand, bend your arm back to your shoulder and slip the coin in the collar of your shirt or sweater. *At the same time,* take the three right-hand fingers and rub them on your elbow as before. When the quarter is safely tucked into your collar, throw both your hands forward, open your fingers, and yell "Presto!" Well, you warned your audience to keep their eyes on the quarter! Anyone who did should have caught the trick. But most people will be distracted by the dropping coin and won't notice that you switched the pick-up hand. That's magic!

Other magic tricks need special equipment or "props" to help create an illusion. Illusionists are magicians who work on a stage, farther away from a large audience. In these stunts, people may be made to "disappear" or be "changed" into other people or animals. (Did you think the lady in the box was really cut in half at the last magic show you saw?)

Natural illusions, such as *mirages,* can also occur—usually in wide, open spaces like deserts and oceans. For a mirage to happen, a layer of warm air must be sandwiched between two layers of cooler air. Light is refracted (bent) each time it enters or leaves one of these layers. This refraction can create the illusion of lakes and trees—even a whole city! The "lake" may even appear to have waves rippling across its surface. This is caused by the shifting of the different layers of air.

Since mirages are made by actual rays of light, it is possible for several people to see the same mirage. In fact, mirages can be photographed. So the next hot summer day, keep your eyes open for mirages—like the "puddles" you see ahead of you when you are riding on the highway. Don't let your eyes trick you into trying to splash around in one!

ANTS IN YOUR PLANTS

ANTS IN YOUR PLANTS

Each Ants in Your Plants experiment takes about 10 minutes to set up. The amount of time you spend observing the ants is up to you.

YOU WILL NEED

1 Spoonful of tuna
or meat scraps
1 Spoonful of honey
or maple syrup
1 Piece of fruit
or other picnic food
Jar lid, bowl of water, stick

What would the ants eat if you invited them to your picnic? Would they go for the salami sandwich first or head for the watermelon? Invite the ants in your plants (or lawn or park) to a picnic just for them. You may be surprised by their special ant antics.

BEFORE YOU START:
Can an Ant Make You Say Uncle?

Ants belong to the family Formicidae of the order Hymenoptera. This may not seem important until you realize that bees and wasps belong to the same order. Some ants have stings and some can spray poison from the end of the abdomen. Most ants have strong jaws and will bite to defend themselves and their nests.

The bite of one tiny ant can hurt a little. The bites of many ants can hurt a lot. That is why you must be very careful when dealing with ants.

■ When ant hill hunting, always wear protective clothing, such as long pants tucked into high boots. Heavy socks will help.

■ Avoid disturbing ant colonies. If you step in the wrong place, your legs will be covered with ants before you know it.

■ Don't try to pick ants up with your fingers. Use a stick or other utensil to keep a healthy distance between you and the ants.

EXPERIMENT No. 1:
HILL HUNT

1 Go on a hill hunt on your lawn or in your park to find an ant hill. An ant hill looks like a small pile of dirt with a hole right in the middle of it. Sometimes when you spot an ant walking through the grass, you can follow it back to its hill. If you can't find an ant hill, find a bare spot on the ground to do your ant hunting.

2 Put a small spoonful of tuna or meat scraps about one foot away from the hill or the center of the bare spot on the ground. A spoonful of cat food will work, too.

3 Pour a small spoonful of honey or syrup on a leaf. Lay the leaf about one foot away from the ant hill or bare spot. Make sure it is also at least one foot away from the tuna or meat.

4 Set out fruit and any other kinds of picnic food you think the ants in your plants might enjoy. Put each kind of food one foot away from the ant hill or bare spot and one foot away from any other kind of food.

5 Watch carefully. Which kind of food do the ants find first? With each food placed one foot from the ant hill, the ants will choose their favorite foods, not just the ones closest to them. Some ants like meaty food like tuna and insects. Other ants would rather eat sweet foods like honey and plant juices.

Do the ants eat the food on the spot or do they carry it back to the hill? Do they work alone or do they help each other? As you observe the ants, you may notice that when an ant finds food, it runs back to the hill to "tell" the others. As it runs, it leaves a trail that other ants in the hill can smell. The ants find the food by smelling their way along the trail. You can use these "smell trails" to make more observations about ants.

EXPERIMENT No. 2: ANT ANTICS

6 Find a place where several ants are going back and forth on the same smell trail. Lay a stick across the smell trail and watch what happens.

7 Carefully scoop up a few ants. A paper plate may be useful for this purpose. Gently put the ants on the inside of a jar lid. Float the lid in a bowl of water. The ants can't escape because they can't swim. You can also keep them from wandering by coating the edge of the jar with petroleum jelly or grease.

8 Place a piece of tuna or meat in the jar lid with the ants. Watch to see if they eat it. Check if there is more than one kind of ant on the lid. Do the different ants fight over the food? Which type wins?

After making your ant observations, go back and look at the pieces of food and sweets you placed around the ant hill. Are they still there?

VARIATIONS

- Some ants come out in the open to eat only at night. Perform the Hill Hunt experiment at night. Use a flashlight to check for new kinds of ants feasting on your food bait.
- Can your mother smell as well as an ant? How about the other members of your family? Make a smell trail on your lawn or in your park and let them try. Put several spoonfuls of extract—peppermint, coconut or almond—in a spray bottle or mister. Fill the bottle with water. Make a trail by spraying the extract every six feet in a line. See if your family can sniff its way from beginning to end.

HOW TO BE AN ANT FARMER

You can buy an ant farm kit in a pet store or through a catalog, but it's much more fun to make your own. Get a large, clean glass jar and fill it with soil. Set it in a pan of water to keep strays from wandering off the farm. Lay a piece of cardboard with a small breathing hole in it on top of the jar. Coat the rim of the cardboard with petroleum jelly or grease.

Place the ants from your lawn or park in the jar with care. Start out with at least two dozen. Try to add a queen

ant to your farm so you can watch the whole life cycle. The farm won't last long without a queen. You may find it worthwhile to buy ant eggs to populate your farm. Feed the ants dead insects and corn syrup. Give them a wet sponge or cotton ball to suck on for water. Soon the ants will dig tunnels and rooms as they go to and fro along their smell trails.

AFTERWORDS

Ants are just about everywhere. You can prove this fact whenever you go on a picnic. Just find a nice picnic spot without an ant in sight, spread out your blanket, unpack your picnic basket, and suddenly it's ant city. Where do the ants come from? They probably were living in the ground right under your picnic basket. Many kinds of ants live underground in ant cities called colonies. Each colony is ruled by at least one extra large queen ant. A queen ant can live as long as fifteen years. Her job is to lay eggs to produce more ants for the colony.

Ant colonies can contain thousands of ants. Most of these ants are worker ants. The workers build and maintain nests and hunt for food for the queen, for the young ants, and sometimes for soldier ants that defend the nest from invaders. Worker ants carry this extra food in their second stomachs. Their first stomachs are used for their own food.

The way some ants collect food is fascinating and almost human. One type of ant keeps tiny insects called aphids like farmers keep cows. Aphids suck the sweet juice or honeydew from plants. The ants like to eat honeydew. To get the honeydew from the aphids, the ants "milk" them by gently rubbing them with their feelers. In the winter, the ants bring the aphids into the nest.

Other kinds of ant farmers cut pieces out of leaves. They chew the pieces until they are soft, then spread them out underground. Soon, a fungus grows on the chewed leaves. The other ants eat the fungus.

Some worker ants in the desert use their second stomachs to store honeydew. These workers are fed any extra honeydew. They get so fat, they can hardly move. They just hang in the nest like little honeypots. Other ants take honeydew from the honeypot ants.

While worker ants feed the colony, soldier ants protect it. One type of tree-living ant guards the nest by grouping their heads together like a door. They open the door to let colony members in and out, and shut the door to keep other ants out.

Tailor ants in Asia, Africa, and Australia sew leaves together to make nests. The ants living in bullhorn acacia trees in Mexico live in the trees' thorns. The tree makes nectar for the ants to eat. The ants protect the tree by getting rid of insects and animals that try to eat it. The tree couldn't live without ants. Tree-living ants are just one variety of ants that don't live underground. Ants live in nests in dead wood, in living plant tissue, or in papery nests attached to twigs or rocks. Ants also may invade buildings or ships.

The driver ants of Africa have no permanent nests at all. They are almost always on the move. When millions of these ants are on the march, they eat everything from insects, birds, and small animals to large animals such as elephants. They have even been known to eat people who could not get away. When driver ants are on the march, people don't worry about ants spoiling their picnics. They worry about becoming the main course at an ant picnic.

PIZZA GEOGRAPHY

PIZZA GEOGRAPHY

Pizza Geography will take about 90 minutes to complete and eat using ready-made pizza dough. Add about 2½ hours to make the crust from scratch.

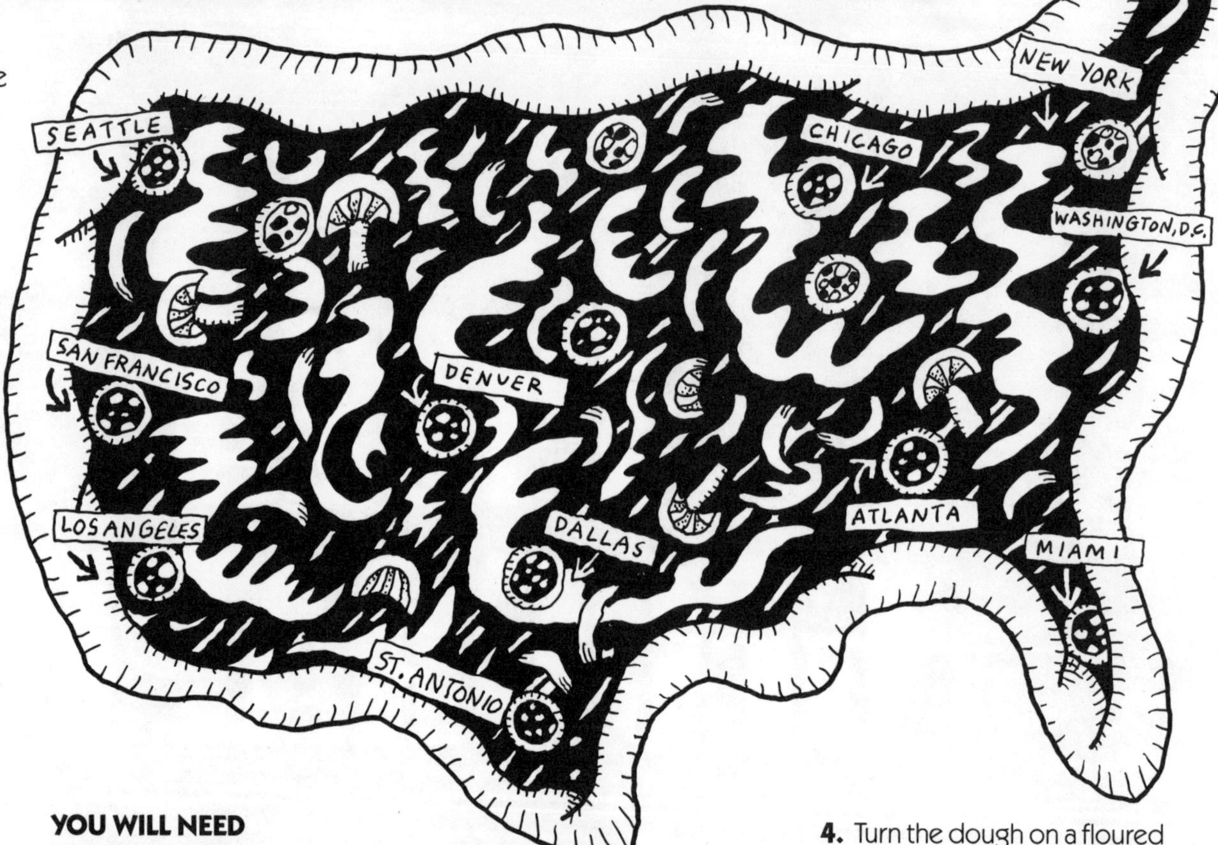

YOU WILL NEED FOR ONE PIZZA

8 Ounces tomato or pizza sauce
½ Pound grated mozzarella cheese
Sliced vegetables (onions, peppers, etc.)
1 Cookie sheet (11″ × 15″)
Spices (oregano, garlic, basil, etc.)
Sliced cooked sausage or pepperoni
Any other topping you would like to try
Oil, oven
Rolling pin or drinking glass
Atlas or other book of maps
Camera (optional)
Ready-made dough or ingredients at right.

YOU WILL NEED FOR PIZZA CRUST

¼-Ounce package dry yeast
1⅓ Cups warm water
2 Tablespoons oil, 1 teaspoon salt
1 Tablespoon honey or sugar
4 Cups flour
1 Freezer container
Mixing bowl, clean dish towel

1. In a mixing bowl, dissolve the yeast in warm water.
2. Add oil, salt, and honey *or* sugar. Mix well and let stand for five minutes.
3. Add the flour one cup at a time. Mix between additions. Add flour until dough is elastic, but sticky.
4. Turn the dough on a floured board or counter and knead for a few minutes. Kneading means folding and pressing the dough. Check a cookbook for more details on kneading, if needed.
5. Oil a mixing bowl and flop the dough into it so the dough gets coated with oil. Cover the bowl with a towel.

Let the dough rise in a warm place for 1 to 1½ hours or until it is double in size.

Yield: Two 11″ × 15″ pizza crusts. (You can freeze one-half of the dough after it rises. Place it in an airtight freezer container. When you want to use it, just defrost the dough, punch it down, spread it on an oiled cookie sheet, and add toppings.)

A map can be more than just numbers and lines, and a pizza can be more than just topping and crust. Put a map and pizza together for a fun dinner that you can design yourself.

1 Make the pizza dough. If you are in a hurry, buy a ready-made pizza dough or crust mix. While the crust dough is rising, look through an atlas (a book of maps). Decide what type of map you are going to make (weather, population, major rivers, etc.).

2 Preheat the oven to 425 °F. Coat a cookie sheet with oil. Plop the dough onto the cookie sheet. Wet your hands with oil or water so the dough won't stick to them as much.

3 Squeeze and press the dough onto the cookie sheet. Use a floured rolling pin or glass to help you flatten it out.

4 Use an atlas to help you form the dough into a map of the U.S.A. or another country. Make a ridge of dough along the edge to keep sauce and toppings from dripping. Spread tomato sauce over the dough.

■ Add toppings to mark features: mountain ranges and rivers; your state, state capital, and towns where friends and relatives live; favorite vacation spots; historic landmarks; major industries or crops or other important or special spots.

5 Photograph or draw your pizza map. This will be your "before-baking" picture. Once the map goes in the oven, the cheese will melt and spread. Your map will look different after it has been baked.

6 Bake the pizza map for 20 to 30 minutes. The baking time will depend upon your oven and the thickness of your crust. Keep checking. The pizza is done when the cheese melts and the crust is brown.

■ While the pizza bakes, play Geography. What is the capital of Delaware? Where is Glacier National Park? Think up questions to ask each other. The answers can be found in the atlas.

7 When the pizza is done, take it out of the oven and let it cool. Photograph or draw your "after-baking" picture. Now eat the pizza.

AFTERWORDS

If your idea of geography is memorizing the names of the Great Lakes, then you probably think geography is boring. In fact, geography is a fascinating subject, because it covers the whole world. The word *geography* even means "writing about and describing the Earth." Geography includes both natural features of the Earth, such as rivers and mountains, and man-made features, such as bridges and cities. It helps describe the Earth from ancient times into the future. Geography is concerned with all of the forces that change the Earth, from hurricanes that wipe away coastlines to wars that wipe away national boundaries. But geography does not stop on Earth. Geography has also conquered space with maps of our solar system and beyond.

Geography covers such a wide area that no one can handle all that needs to be done. Here are some of the special jobs that geographers do:

- Biogeographers study the living things on Earth and map their locations. Their maps show jungles and grasslands and rain forests and prairies, and what animals live in each environment. Biogeographers must know life sciences and earth sciences as well as geography.
- Demographers are population specialists. They study how and why human populations change. For example, they would study why people seem to be leaving cities to live in small towns. Their maps would show these changes.
- Medical geographers put contagious diseases on the map. They help decide who can travel where, and what shots may be needed before going.
- Economic geographers look at how people earn money and how that affects where they live. They study people whose farm animals graze on grass; they also study automobile workers whose jobs depend upon industry.

A geographer's job is to describe changes on the Earth's surface, but it is even more than that. He or she must often attempt to explain why the changes occur and to predict what will happen in the future.

Geography is a lot more than a game about state capitals.

SPINY STUNTS

SPINY STUNTS

Spiny Stunts takes about 20 minutes to complete. It is most successful when done during the cactus' growing season—spring or summer.

YOU WILL NEED

2 Small, column-like cacti of about the same diameter
Clean, sharp knife or single-edged razor blade
Alcohol, shoelaces or yarn, cotton
Newspaper or paper towel

It's easy to make new plants grow from two old ones—just graft them together! Grafting is a simple way of growing two similar plants from the same set of roots. Plant growers use grafting to make new varieties of plants from old varieties.

You can make your own original cactus combinations by grafting a tall cactus top to a short cactus bottom, a bald top to a hairy bottom, or a flat top to a round bottom—or try your own Spiny Stunts!

BEFORE YOU START: When is a Cactus Not a Cactus?

To do successful Spiny Stunts, both plants that you use must be members of the cactus family. Cacti have spines, but not every spiny plant is a cactus. Some members of the Euphorbia (you-for-bee-uh) and milkweed families that live in the desert are also spiny and can look very much like cacti.

To tell the difference between a real cactus and an impostor, look at the plant's spines. Cacti have cushiony bud-like areas on their stems from which spines, flowers, and hair grow. Spines grow *only* from these buds. In milkweed and Euphorbia plants, spines grow directly from the stem instead of from buds.

A second way to identify a cactus impostor is to stick it with a pin. If a milky-white liquid oozes from the plant, it is not a cactus, but a Euphorbia. Be careful—the milky Euphorbia liquid can irritate your skin and eyes. (If you accidentally come in contact with it, wash the area thoroughly with soap and water.)

KEEPING THINGS CLEAN

Cactus grafting is an operation that you can do on plants. Just as in any operation, it is important to keep everything as clean as you can. Otherwise, germs will get into the cacti's wounds and the cacti won't heal properly.

1 Dip your knife in alcohol to sterilize it. Let it air dry. Be careful not to touch the blade of the knife again before grafting your cacti. Wrap a paper towel or a piece of newspaper around

each cactus to protect your fingers from the cactus spines. Find a section of each cactus that is about the same thickness. Cut off the top of each cactus with a horizontal cut at that spot. Remember not to touch the open wounds—it's important to keep the cut surfaces germ-free.

2 Look closely at the insides of the two cacti. You will be able to see three layers of tissue. The outer green tissue produces food for the plant. The lighter green tissue in the second layer stores water. The white, inner "pith" contains the growth rings of the cactus. These growth rings make the cactus grow outward from its center.

3 Switch the tops of the two cacti. Put the top of one cactus on the base of the other, and vice versa. Line up the piths of the two cacti as closely as you can. The piths must eventually grow together for the cacti to keep living.

4 With a cotton ball to protect your fingers, push the top cacti firmly onto the bottom cacti. This will remove any air bubbles

between the two. Use a cotton ball to pad the top of the cacti. Tie the cacti in place with shoelaces or yarn.

5 Place your new plants in a sunny window. Keep the cacti warm and the grafts dry and clean. After 3 to 4 weeks, remove the yarn from your cacti and gently wiggle their tops to make sure that the grafts have healed completely. (If the grafts have not healed, make new cuts and start again.) Then enjoy your original houseplant!

CACTUS CARE

Cacti grow at different rates during the winter and summer. For this reason, your cactus will require special kinds of care during different times of the year. A cactus' growing period occurs in the spring and summer. During these months, place your cacti in full sunlight—outdoors if possible. Water your cacti whenever the soil is dry. A small cactus may need water every few days, while a larger cactus will need water less often. When in doubt, wait another day before watering. It's better to under-water your cacti than to over-water them.

During the winter, a cactus enters a rest, or dormant, period. It is important that your cacti have an undisturbed dormant period to get ready for the next growing season. Place your cacti in a cooler location; 55 to 60°F is a good temperature for the winter. Keep your cacti dry; they will be able to store enough water to go 3 to 4 weeks between waterings. Soon after the dormant period, your cacti may begin to flower. Look for buds in February or March.

AFTERWORDS

Grafting cacti can be a sticky business—a real spiny stunt. But the cactus family is exceptionally tolerant of the grafting process. Almost all kinds of cacti can be grafted as long as they are healthy, the cuts are kept clean, and the piths line up closely.

Cactus grafts are successful because all cacti have an active cambium layer inside their piths. The cambium layer is thin tissue that produces new cells that control the cactus' sideward, or lateral, growth. These new cells pass food and water up and down the entire plant. Because it is possible to connect the cambium layer of one cactus to another, cactus grafts "take" readily and the rootstock of a healthy plant can pass food to its more fragile grafted top.

Cactus grafting is fun and produces an unlimited number of "living sculptures." However, horticulturists (people who grow fruits, vegetables, flowers, and plants) also graft cacti for practical reasons.

Occasionally, a cactus grower will find a plant that is unlike any of the thousands produced from the same parent plants. A genetic change makes the difference. These cacti are rare and often beautiful. Unfortunately, they are also often weak and cannot live long on their own. To make sure the most beautiful or unusual part of the cactus survives, a cactus grower grafts it to a normal cactus with a hardy set of roots, or rootstock.

A striking example of using grafting to save an unusual plant is in the case of albinos. Albino cacti lack the green chlorophyll they need to produce their own food. These "sports" can be bright red, pink, or yellow, instead of green. Without chlorophyll, the albino cacti will die. Cactus growers provide these cacti with a constant food source by grafting the albinos on to healthy green rootstocks.

Cacti aren't the only plants that are grafted. Many fruits are grown from grafted plants. Apple trees are grafted to produce better varieties of apples. Sometimes apple growers graft branches from tall apple trees onto short apple trees. Then they can reach the fruit without using ladders.

Every time you eat a seedless orange, you are eating fruit from a grafted plant. Seedless oranges have no seeds to plant to grow new trees. Orange growers grow more "seedless orange trees" by grafting the fruit-making branches onto the roots of other orange trees.

You can't graft every kind of plant together, but you can make some funny combinations. Sometimes fruit growers can grow a couple of different fruits on the same tree. For example, they can graft a branch from an apricot tree onto a plum tree. The result is plum strange: a tree that grows both apricots *and* plums.

CASTS OF CHARACTERS

CASTS OF CHARACTERS

You can record what you can see and touch by taking pictures or making drawings. But there's another way that's longer lasting, more lifelike—and more fun! The casts can be made in about 20 minutes, plus an hour's drying time.

YOU WILL NEED

Door key
Plasticine or modeling clay
Paring knife
Some newspapers
Empty cardboard milk carton
6 Paper clips
2 Teaspoonfuls petroleum jelly (such as Vaseline)
Plaster of Paris, at least a pound (Buy the cheapest kind you can get. Drugstores sell small amounts; building supply stores sell large sacks.)
Empty half-gallon plastic ice-cream carton
Small, stiff brush (An old toothbrush will do.)
Stirring stick or wooden ruler, spoon
Paper towels and dish detergent
Paint, any kind: poster, watercolor, spray
Wire coat hangers

Suppose you are hiking in the Pacific Northwest and find some footprints—footprints of the legendary *Sasquatch*, "Bigfoot," the wildman of the Rockies. How can you bring back a footprint to prove to the rest of the world that you found him? You *could* take photographs, but here's a better way.

1 First, practice your technique! Start out with something smaller than a footprint—like a door key. Roll out a pad of Plasticine or clay about as big as your palm and about ½" thick. Put it on a pad of newspapers on the table. Use a paring knife to cut it into a round cookie shape.

2 Lay your key flat on the center of the Plasticine and press down hard. Make the top of the key level with the top of the Plasticine. Now, very carefully lift out the key.

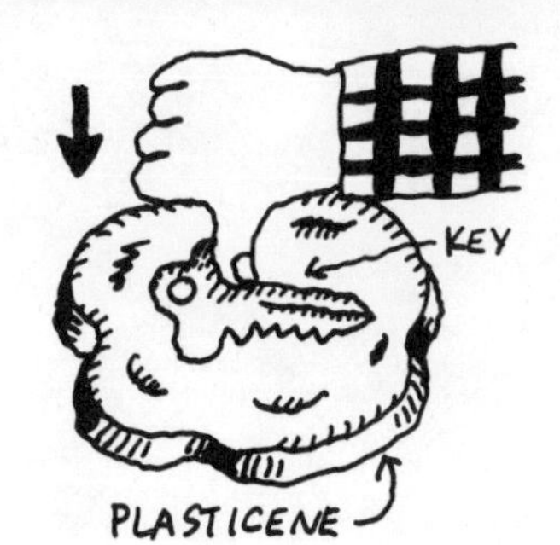

3 Cut a cardboard milk carton into a strip about 1½" wide and as long as you can make it. Wrap this strip around the edge of your "cookie," with the waxy side against the cookie. If the strip is not long enough, cut another one and overlap them. Use paper clips greased with petroleum jelly to join the strips. Try to make the top edges level.

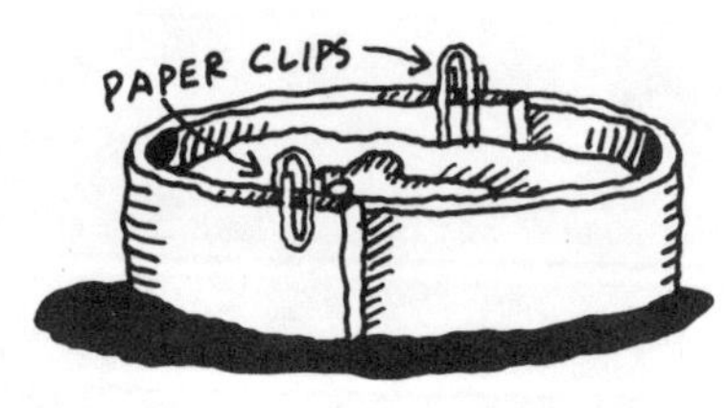

4 Put one cup of water in an empty plastic ice-cream carton. Stir in one cup of plaster of Paris. Stir until it is well mixed and starts to look like cake frosting. Work quickly. Plaster of Paris hardens in a few minutes. **Do not** try to add more water if it is too thick. (You *can* add more plaster.) **Caution: Do not put your hands or fingers in the plaster mix and leave them there "to see what happens"! And do not dump any of the wet plaster down a drain. Leave it in the carton.**

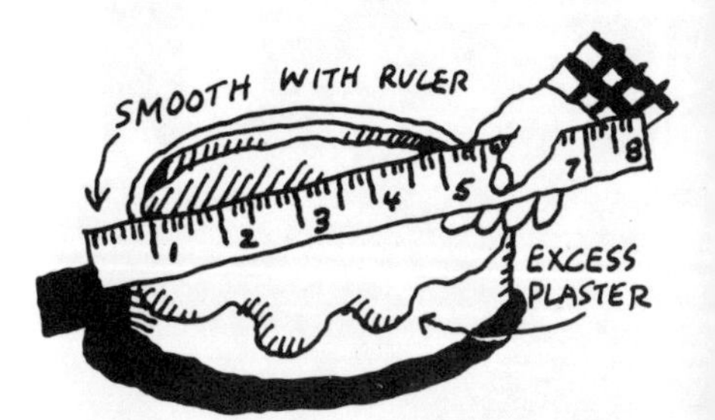

5 Pour your plaster mix on top of the Plasticine. Fill it up to the edge of the cardboard. Use the edge of a ruler or knife to smooth off the top. Wipe off the knife or ruler with a paper towel.

6 Bend a clean (ungreased) paper clip into a loop. Stick the ends of the wire into the wet plaster in your mold. This loop becomes a wall hanger.

7 After a few minutes, the water that may have been on top of the plaster disappears. Use one end of a paper clip to write your name and the date in the drying plaster. Let everything sit for a half hour. Touch the surface of the plaster to see if it has hardened. Carefully unwrap the cardboard strips and lift off the plaster. This is your *negative mold.*

8 Smear the whole bulging side of your negative mold with a *thin* layer of petroleum jelly. Use a small, stiff brush to help work it in and out of the grooves and holes in the "key." Put another greased cardboard collar around your negative mold, waxed side inside. Pour more plaster to make your *positive mold.* Label it and put in a hanger. After an hour, separate your positive and negative molds **carefully!**

9 Two hours after pouring, wash your molds under running water. Use detergent to get rid of any grease. Once the molds are dry, you can paint them any color. Metallic paints like gold or silver look really great!

10 Now go outside and make casts of tracks: footprints of people or other animals, tire-tread marks, whatever looks interesting. When you find a track you would like to cast, clean any bits of leaves or other junk out of it first. If the soil around the track is crumbly, you can spray the track with hair spray. This will harden and hold little pieces in place. Put a greased cardboard collar around the track and pour in your plaster. **Remember:** You can't pre-mix the plaster before you start, so you will have to carry your supplies with you separately.

If you want to make a large cast, bigger than a dinner plate, you will need some reinforcement. Pour in half of your mix and then crisscross several pieces of coat-hanger wire on top of the wet plaster. (Cut the wire with pliers or bend it back and forth until it breaks.) Then pour in the rest of your mix. You might want to leave a loop of wire sticking out for a hanger.

AFTERWORDS

Plaster is made from the mineral *gypsum*. Gypsum looks like rock and is made up of *billions* of crystals. Pieces of gypsum are heated in a fire until about three fourths of the water in the rock evaporates. This breaks down three fourths of the crystals in the gypsum and the rock turns to a powder. When water is added to the powdered rock, presto! The crystals re-form and the powder hardens once again. Instant rock! As the crystals re-form, heat is given off. You can feel this heat while your casts are hardening.

Plaster has been used for about 3,000 years. Our modern product is called plaster of Paris because the soil around Paris in France has much gypsum. Fine-quality plaster has been made in Paris for hundreds of years. But your stuff was probably made in the U.S.A. Where? Check the label.

Police officers often use plaster to make casts—particularly of tire-tread marks and footprints found at the scene of a crime. Then they can bring their outdoor evidence indoors, to the courtroom. But folks in the high Rocky Mountains of the Pacific Northwest (Washington and Oregon) have found some curious footprints, too. People have gone into the wild areas and brought back casts of huge footprints. These prints *look* like they were made by humans but are too big to belong to any "normal" man. These are supposed to be prints of prehistoric humans—the Sasquatch, or Bigfoot, of the Rockies.

Some people believe these oversized, human-like creatures exist. Some people even claim to have seen them—describing them as about 7 feet or taller, with long, dark hair all over their bodies. Other people say there is no such thing as Bigfoot. They say it is all a fake and the footprints have been made by someone wearing oversized shoes with soles shaped like feet.

However, since tracks have been found in very wild places, where few people go, whoever is playing the joke is going through a lot of trouble. And the trick wouldn't be that easy to pull off. Look at your footprints in sand or snow. Now stand on a board bigger than your foot. Where do you sink in the most? You would need to carry a very heavy load to sink the board as far in as your foot alone would sink. And while you were carrying the load, you would have to take strides about twice as long as normal!

Still, if it *is* a hoax, then it has gone on for a long time. There are stories about the capture of a Bigfoot in the Canadian Rockies about 100 years ago. Other countries also have stories and legends about wild humans. The natives of the Himalayan mountains between Nepal and India speak of their "Yeti," or Abominable Snowman. Maybe these creatures exist, maybe they don't. But no one has managed to capture one recently.

What do you think? What will *you* accept as "evidence"? Until you decide, don't go walking in the Rockies alone at night!

DRAWING MACHINE

DRAWING MACHINE

Here's a simple "drawing machine" you can make in one hour. Use it to copy a picture in any size you want!

YOU WILL NEED

Corrugated cardboard, about 20″ × 10″
2 or 3 Pieces of corrugated cardboard or 1 piece of plywood, each about 20″ square
White drawing paper or shelf paper
3 to 6 Paper clasps (small brass pins with rounded heads attached to two metal strips that can be bent back; strips must be at least 1″ long)
1 Thumbtack or pushpin
An awl or hand drill, tape, cutting knife or scissors, ruler
1 Pencil and 1 felt-tip pen or 2 soft-lead pencils

It's easy to *trace* a picture — but have you ever tried to copy one and make it larger or smaller? You may start off okay, but then maybe a line is too long or too short and you have to start all over again. With this drawing machine called a *pantograph,* you can have fun tracing a design or picture and watching a smaller or larger copy appear on the paper too!

1 Cut the corrugated 20″ × 10″ cardboard (from the top or bottom of a box) into 4 strips 11″ × 1¼″ and 4 strips 7″ × 1¼″. (It's okay if you make them a little wider.) Be careful with the cutting knife!

2 Divide the 11″ strips and the 7″ strips into sets of two strips each. On each strip, put a pencil mark ½″ in from one end; then put a mark every 2″, beginning from the first mark. You should have 6 marks on the 11″ strips, and 4 on the 7″ strips.

3 With an awl or hand drill make a hole at each mark on each strip. Then tightly tape each set of strips together, one on top of the other, to make 4 separate "drawing arms." Be sure you don't tape over the holes!

4 With a paper clasp, fasten the two longer strips together at the ½″ mark (Point A in the illustration). Fasten the other two 7″ strips to the long strips at Points B and C. That's your pantograph!

5 To make your drawing board, tape the large square pieces of corrugated cardboard together, one on top of the other; or just use the piece of plywood. Cover the drawing board over with paper to make a clean, flat working surface. With a thumbtack or pushpin, firmly anchor your pantograph to the drawing board at Point D.

6 Push your pencils into the holes at Points E and F. Be sure the pencil at Point E holds the two drawing arms together. If you have trouble, use a little tape to hold the pencil to the lower arm. (Don't tape the two arms together; they must be able to move.)

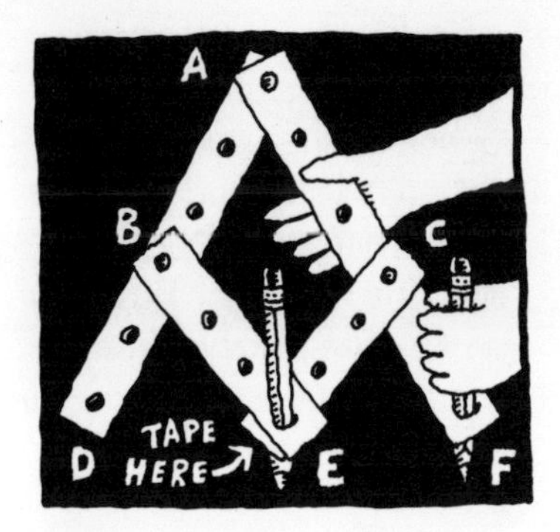

7 Now you're ready to "draw double." Set your paper on the drawing board, under the pantograph. Tape each corner of the paper to prevent it from shifting. If you use the pencil at Point E to draw your design, the pencil at Point F will make a large copy of your design. If you use the pencil at Point F, the other pencil will make a smaller copy!

VARIATIONS

■ You can change the copy of your drawing just by adjusting the connections of the drawing arms. Experiment by moving the connection at Point B one hole closer to D. What happens to the copy of your drawing? It may look distorted — like something in a funhouse mirror! Next try attaching the short arms to different points on the long arms. Move the pencils to different points. Draw first with one pencil and then with the other. Can you figure out what happens when you make the changes?

■ You can also extend, or lengthen, your pantograph. (But then you will certainly need a larger drawing board!) You can add more arms too; cut a few extra 11″ strips of cardboard in sets of twos taped together to make stiff drawing arms. Make the same holes as you made in the first set. Then attach your new arms and make two or three copies, adding more pencils.

■ The copies may come out too light if you use a lead pencil; why not try a colored felt-tip marker, the size of a pencil, on the copying arm instead?

■ If you have a favorite comic book character, you can use your pantograph to trace it in a different size. Put the picture on the drawing board, under your paper, and trace along the outline with the pencil at Point F. Is your copy larger or smaller than the original?

■ If you have a hand drill and a 3⁄16″ bit, you can use thin wooden strips of lath, instead of cardboard, for the drawing arms. You can even use cheap wooden rulers that you buy in the dimestore. You'll need at least three (cut one in half), but four would be ideal. Drill

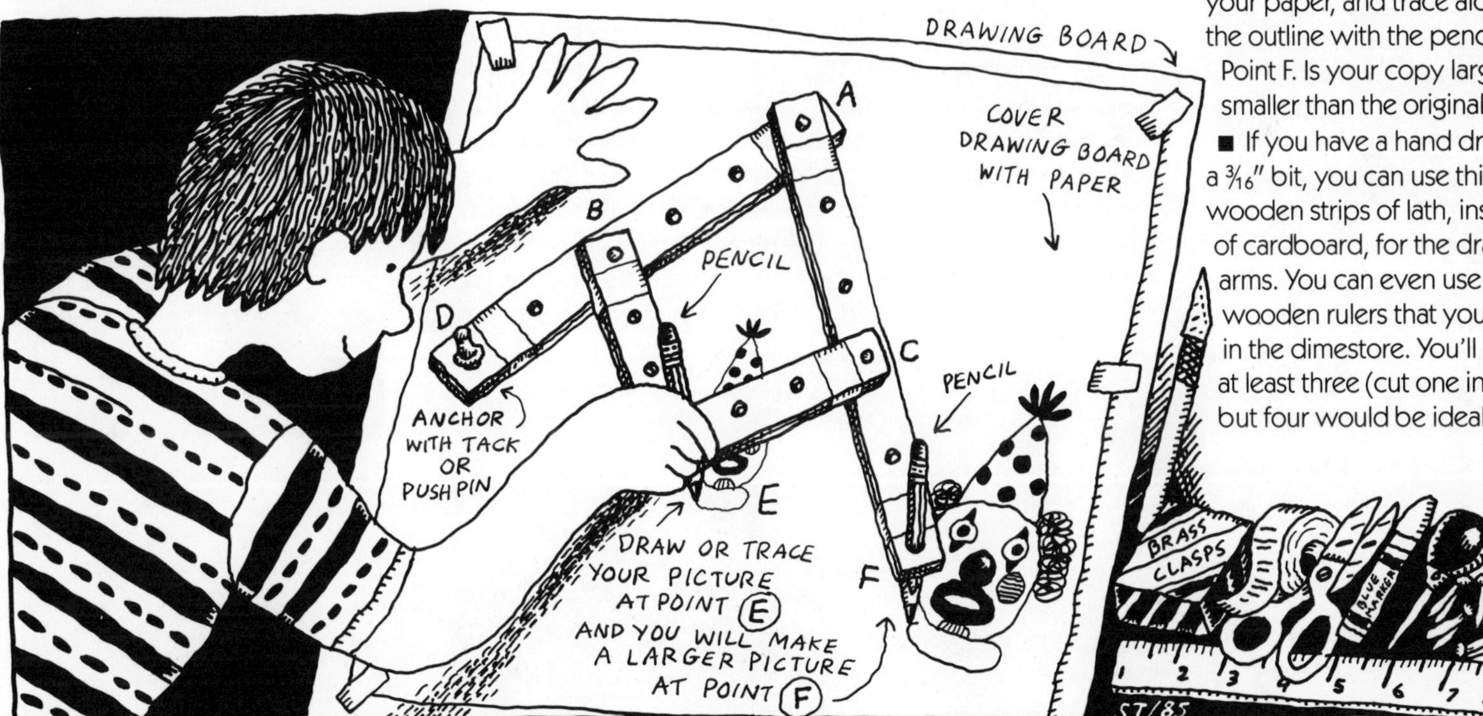

your holes and use paper clasps that are at least 1″ or 1¼″ long to attach the drawing arms.

Copying machines in an office may make copies faster, but it's much more fun to make your own pantograph and watch it make copies as you draw!

AFTERWORDS

Do your eyes ever play tricks on you? Perhaps — when a magician seems to pull a coin out of the air. But a pantograph plays no tricks on you when it enlarges or reduces a picture you draw or trace.

This change in size depends on the *ratio* of the distances between the tracing instrument (your pencil) and the drawing, or copying, instrument (the second pencil or felt-tip pen) to the pivot points (the points where the arms of the pantograph are attached). If the distance increases evenly — that is, if points B and C in the drawing above are changed by the same amount — the copy will get bigger. Some pantographs will enlarge a drawing as much as 15 times its original size — a ratio of 15 to 1. By reversing the positions of the tracing and drawing instruments, you can make the picture smaller.

Interesting distortions can be made by changing the pivot points so that the arms are no longer parallel. For example, adjust the pivot point at A. Put the paper clasp through the first hole of one arm and the second or third hole of the other arm. If the tracing pencil draws a circle, the copying pencil will produce an oval or elipse. Trace a square and your copy will be a diamond!

Another way to enlarge a picture? Look at it through a magnifying glass! The glass is a double-convex lens, curved out on both sides. It too relies on the relationship of distance to a point of contact, or *focal point*. But your eyes play a key part. We see an object because light rays bounce off of that object and into our eye (see diagram). As they pass through the *cornea,* then the *iris* (the colored part of your eye), the *pupil* (the black part),

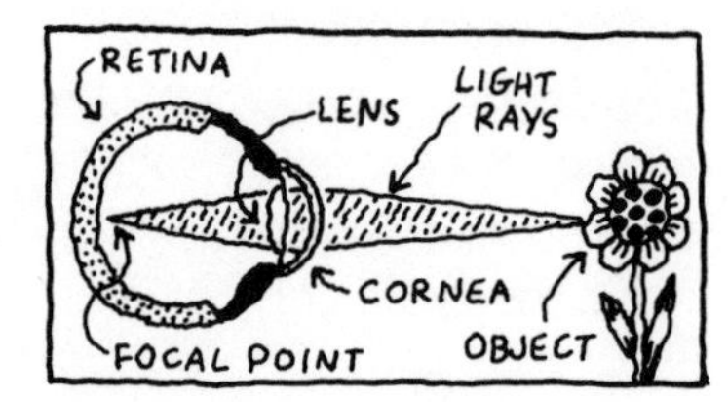

and the lens, the light rays are bent and brought together at the focal point of the *retina.* Strange as it seems, the image that is focused on the retina is *upside down* and *flat!* But by means of the many nerves to the brain, the image is processed into the correct image that we "see."

When we look through a magnifying glass, light rays bounce off the object, pass through the lens and are bent inward to meet at a point called the *focus* (see illustration #2). The distance from the center of the lens to the focus is the *focal length.* If the object (a word on a page, a leaf, or some other small item)

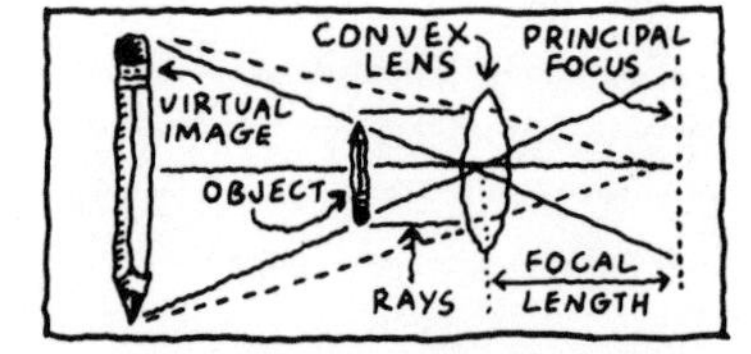

is less than one focal length from the lens, you will see an image that is right side up and larger than the original object.

Enlarging pictures with a pantograph or a magnifying glass can be fun. They work in different ways, but one thing never changes: The way your eyes work when they're *not* playing tricks on you!

HIGH FLIERS

HIGH FLIERS

Hats off to windy days! A blast of wind can give your spirits a lift. It only takes 20 minutes to experiment with a few low fliers, but you'll need 2 hours to build a high-flying kite of your own.

YOU WILL NEED

Hair blow-dryer
Several 8½" × 11" pieces of paper
Scissors, ruler, pencil
Kite line (20-pound test)
1 Sheet of 27" × 27" tissue paper, airplane paper, or crepe paper
3 Hardwood or softwood strips, each ¼" square × 24" long
Elmer's white glue
Rubber cement
Sharp utility knife or small saw
Paints for decoration (optional)
1 Small plastic ring (available in the drapery section of a department store)
Needle with large eye

Have you ever seen a round kite? Do you think a round kite would fly? There's only one way to find out—experiment! The best kite shapes will create a reaction in the air around them. This reaction is called *lift*. You get lift when the pressure on top of a kite—or bird or plane—is less than the pressure underneath it. In High Fliers, you can "test fly" many flat shapes in your living room to find out which ones will soar. Then make a kite and fly it, to feel the power of the wind in your hands. That should give *you* the biggest lift of all!

TEST FLIGHTS

1 Plug in the blow-dryer near a table. The table is going to be your testing runway. Put a sheet of paper down flat on the runway, three inches from the edge of the table. Hold the blow-dryer several inches above the table and turn it on to the low setting. You might have to move the hair-dryer around a little bit to get the angle right, but very soon the paper should lift off and "fly" across the table.

2 Cut out a circle, a diamond, an oval, a hexagon, and several other shapes from the remaining pieces of paper. Try to make all the shapes about the same size. Test fly each shape on your runway, and try to decide which shape flies best. **Note:** Don't expect the shapes to fly very high. Half an inch off the table would be an excellent flight height!

3 Repeat the experiment, but this time crumple up one or more of the paper shapes first. Do the crumpled pieces lift off the table?

A HIGH-FLYING KITE

4 To get your project "off the ground," you have to make a kite frame. Use a sharp utility knife to cut a shallow (⅛") square notch in the center of two of the hardwood or softwood strips. These strips are called *spars*. (They go from side to side across the kite to give it support.) Cut square notches four inches in from each end of the third wooden strip.

This strip is called the *spine*. (It goes from top to bottom, and is the kite's "backbone.")

5 Now you will need to make V-shaped notches in the ends of the spars and the spine. Later you will be running a piece of kite string around the kite frame through these V-notches, so be sure that the V-notches are going in the right direction. The diagram shows you how the V-notches should look in relation to the square notches you already made.

6 Latch the spars to the spine to form the kite shape shown. To latch them together, first put some Elmer's glue in the square notches and fit them together. Then use a piece of kite line to tie the spars to the spine at the square notches. Tie the string in a crisscross fashion, going around one way and then the other. When you have finished latching, pour a small amount of glue over the strings, and let it dry.

7 Tie a piece of 20-pound-test kite line to one end of the center spine, hooking the line into the V-

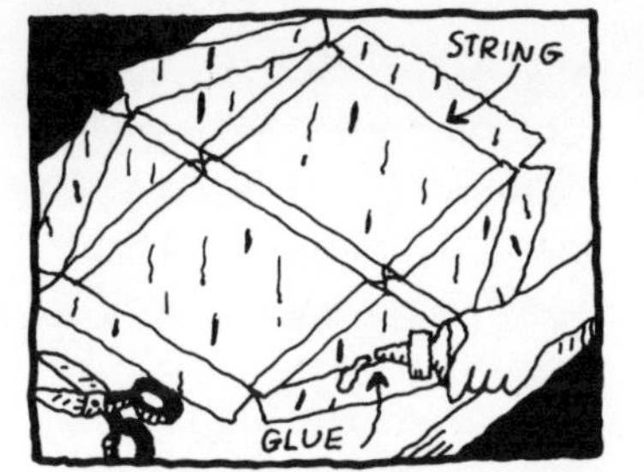

notch. Run the kite line all the way around the frame through the V-notches, pulling it tight but not tight enough to bend or snap the spars. Tie the kite-line ends together when you get back to the beginning, at the center spine. Adjust the frame until it is "squared up" and symmetrical, so that the spars and spine meet in a right (90°) angle.

8 Using the frame as a pattern, cut a kite cover from the 27″ × 27″ piece of paper. Be sure to *allow one inch extra* all around for a "hem." Cut slits at each corner of the cover, where the spars and spine are. Fold each edge over the outline string and glue it in place using rubber cement. Follow the tradi-

tion of kite-makers everywhere by decorating your kite boldly, using *lots* of color.

9 Make a bridle by attaching four pieces of kite line, 12″ each, to the spars as shown in the diagram. You can use a needle with a large eye to do this. The bridle goes on the front of the kite, so you'll have to poke small holes in the cover and stick the line through it to tie the bridle on. You may reinforce the cover with Scotch tape if you want to, but use only a small amount of tape right near the bridle holes. Tie all four bridle lines to a tow ring—a small, lightweight plastic ring like the ones used on drapery and curtain rods. Then tie the rest of the kite line to the tow line... and go fly your kite!

TIPS FOR KITE MAKERS

- Make your kite as light as possible.
- Before constructing your kite, be sure the spars are balanced. Measure to find their exact centers, and then balance each spar on one fingertip. Shave off small pieces if necessary.
- If your kite won't fly, adjust the bridle. Make it shorter or longer. Or attach it somewhere else along the spars. (Cover over the old holes with small pieces of tape.) Keep trying!

AFTERWORDS

For more than 2,000 years, the people of China, Japan, Korea, and many parts of Asia have been flying elaborate kites—most often for pleasure, but sometimes with other purposes as well. Fishermen from China and Japan have been seen standing on shore with a kite flown out to sea. Dangling from the kite line is another string, which falls into the water several hundred feet from shore. When the fishermen feel the kite pulling away, they reel it in and collect a fish at the end of the second line! For sport, Koreans apply a mixture of glue and crushed glass or sand to the kite line near the kite, and engage in kite fighting. The object is to direct one's kite line toward an opponent's kite, cross the kite lines, and cut his line with a quick jerking motion. So proud are the Chinese of their kite-flying traditions that they set aside a holiday in September, called Kite Day. Similarly, the Japanese fly kites on Boy's Day in May, and the Koreans celebrate the first day of the new year with kites flown from morning until night.

Finally, in the 18th century, kites were put to scientific use. Two young Scottish students, Alexander Wilson and Thomas Melville, attached a thermometer to a kite and flew it to test their hypothesis that air is colder at higher altitudes. Three years later, in 1752, Ben Franklin discovered the presence of electricity in the air by flying a kite during a thunderstorm. He had tied the kite line to a key, and tied the key to a silk ribbon, which he held. He was gambling that if the ribbon remained dry, he would not receive an electrical shock. In fact, the electricity traveled down the kite line to the key, and jumped across the space between the key and his hand, giving him a small shock.

In the late 1800s, another Scotsman, Captain B.F.S. Baden-Powell, developed a way to lift a man into the air in a basket attached to six kites flown in tandem. The kites were hexagonal ones, called Levitors. (The kite you made in High Fliers is a Levitor kite!) Baden-Powell's man-lifting kite was soon adopted by the army and used during the Boer War for aerial observations.

Next, William A. Eddy designed the first kite to use a bowed, or bent, spar. The Eddy kite had better lifting qualities than earlier kites, and it was more stable too. When Eddy presented his kite design to the U.S. Weather Bureau in 1894, they immediately saw the potential for lifting meteorological instruments into the sky, and weather stations continued to use kites until more modern methods replaced them.

Perhaps the most dramatic development in the history of kites came from Lawrence Hargrave, an Australian who invented the box kite, which bears his name. The box kite was shaped just like a rectangular box, open at both ends and in the middle of the remaining four sides. Even in rough winds and bad weather conditions, it proved to be an extremely stable, high-flying kite. Then, Orville and Wilbur Wright turned the box kite on its side, twisted the flat surfaces somewhat, and used it as the basis for the wing span on their first manned flying machine!

CABBAGE CAPERS

CABBAGE CAPERS

If you're a kitchen chemist, you can perform the magic trick of changing purple water to red or blue! These Cabbage Capers will only take you 30 minutes, and you can make the cabbage into a salad when you're through.

YOU WILL NEED

½ Red cabbage
Sharp knife
2 Large bowls
Colander or strainer
2 Small drinking glasses
Water
Lemon juice
Baking soda
Several test foods, such as orange juice, apple juice, tomato, vegetable soup broth, egg white, cola, cream of tartar, cocoa, salt water, vinegar, tea, etc.

When foods taste sour, it's usually because they contain some kind of acid. Lemon juice is a perfect example of an acid food—the citric acid gives it a very sour taste. The opposite of an acid is called a *base.* Bases—baking soda, for example—often taste bitter. Foods that are neither acid or base are called *neutral.*

Who cares about acids and bases? You do, when you're sick or have an upset stomach, because acidic foods can irritate your stomach even more. And people who have stomach ulcers also have to watch what they eat; acid can make their ulcers worse. You can find out which foods are acids and which are bases by doing the Cabbage Capers in this colorful experiment!

1 Cut a red cabbage into thin slices and chop the slices into fine pieces, about the size used in cole slaw. Put the chopped cabbage in a large bowl and add 2 cups of very hot tap water. Stir it up and let stand for 15 to 20 minutes, or until the water turns purple. As you can see, *red* cabbage is actually purple, and the water will be a purplish blue.

2 Separate the purple water from the cabbage by pouring the mixture through a strainer or colander into another bowl. This purple water is called an *indicator.* It will change colors to indicate whether various foods are acids or bases. Save the chopped cabbage to make Confetti Coleslaw with the recipe on the next page.

3 Now you're ready to begin doing some kitchen chemistry. Put a tablespoonful of the indicator in a clean glass. Add a few drops of lemon juice. What color did the indicator water change to? Lemon juice is an acid, so now you know what color to expect when you are testing for acids. Don't pour the lemon juice test out yet. You'll need it in the next step.

4 Put a tablespoonful of the indicator in the second glass and add a pinch of baking soda. Baking soda is a base. What color did the indicator change to this time? All bases will have the same effect on the indicator. Now mix the baking soda test together with the lemon juice test. What color did you get this time? Taste the mixture to see what effect the acid and base had on each other.

Caution: It's okay to taste the foods in this experiment, but do not taste the chemicals tested in the variations below. Tasting is not a safe way to test for acids and bases.

5 Rinse out the glasses well, and continue to test some other foods for acid or base. Remember that if the indicator doesn't change color at all, the food is neutral. Use only a tablespoonful of indicator for each test. For dry foods like cocoa, mix a small amount in water before testing it.

VARIATIONS

■ Collect some water from a local stream, river, or lake. Bring it home and prepare a fresh batch of red-cabbage indicator. Test the water. If the

ABRA-
CADABRA
SOUP
VEGETABLE
CREAM of TARTAR
BROTH
TOMATO JUICE
Vinegar
APPLE
S
LEMON JUICE
BAKING SODA
ST/85
BASE
CABBAGE IS AN INDICATOR
WATER
MIX WITH WATER TO TEST
SOUPS, JUICES AND VINEGAR ARE FOODS YOU CAN TEST
ADD TO COLE SLAW
STRAIN THE CABBAGE
LEMON JUICE IS AN ACID

indicator shows that the water has an acidity or basicity, it probably means that certain bugs, plants, and fish cannot survive in those waters.

- Make your own litmus paper by cutting white blotting paper into strips and soaking the strips in a fresh batch of dark purple indicator. After the strips have dried, you can use them to test for acid or base by dipping them into various liquids. Take them with you whenever you are going to be near a fresh water supply, and check up on the environment.
- Test some things that are not foods, like laundry detergent, ammonia, milk of magnesia, soil from your plants or garden, your saliva, and your shampoo. Do you think acids and bases are always good or bad for the environment?

CONFETTI COLESLAW

½ Red cabbage
1 Carrot
1 Green pepper
½ Onion
Lemon juice
½ Cup plain yogurt
½ Cup mayonnaise
½ Teaspoon celery seeds
1 Teaspoon caraway seeds
Salt and pepper to taste

Chop all vegetables into small pieces. Mix the lemon juice, yogurt, and mayonnaise together, and pour over the chopped salad. Add celery seeds, caraway seeds, salt, and pepper ; toss lightly.

AFTERWORDS

In Cabbage Capers, you tested foods for acids and bases. But many substances other than foods are acids and bases (or *alkalis,* as bases are called when they are dissolved in water). You've probably heard of the well-known dangerous acids like hydrochloric acid, which can dissolve metal. But did you know that the poison in a bee sting is a kind of acid too? As with other acids, the bee sting can be neutralized with a common alkali such as baking soda and water. Stinging red ants, car battery fluid, fertilizer, vitamin C, and aspirin all contain acids as well.

Don't put baking soda on a wasp sting, though, because wasp stings are *not* acids—they're alkalis. Some alkalis are just as dangerous as acids, which helps to point out why *all* chemicals should be handled very carefully. Other alkalis include ammonia, Pepto Bismol, the lye used to make soap, and a chemical compound called lithium hydroxide, which was sent along on the Apollo space missions because it has the ability to absorb the carbon dioxide the astronauts exhaled.

The primary characteristics of acids are that they taste sour (but you *should not* taste any chemicals other than foods), they react chemically with metals to produce hydrogen, they change the color of blue litmus paper to red, and they neutralize bases. Bases, on the other hand, taste bitter (again, don't taste chemicals!), they feel slippery or soapy, turn red litmus paper blue, and neutralize acids.

When you made a batch of red cabbage indicator and used it to test for acids and bases, you made one big assumption about the water used in the experiment. You assumed that the water was neutral, not acid or base. If you tested treated water from a city or town water supply, that assumption is probably correct. But although water is considered to be neutral in the world of chemistry, the water you drink doesn't always start out that way.

Imagine the waters that come from upland rivers or streams, or waters that started out as melted snow at the top of a mountain. Fresh mountain water is practically a synonym for good-tasting water, clean and pure and unpolluted. But are mountain waters neutral? Not necessarily. They often have a slightly acid content, due to the small plants and animals in a river or stream that produce acids as they eat, die, and decay. These acids are not by themselves a sign of pollution.

As the mountain water runs downhill, through the ground and over a great number of limestone rocks, it picks up minerals along the way that make the water more base. Water with a high mineral content is often called "hard" water, and it never tastes as good as the "soft" water from farther uphill. Hard water isn't necessarily polluted, either. As it moves into the big rivers near heavily populated areas, the chances of pollution increase. That's because untreated or inadequately treated sewage water is sometimes put back into the rivers. In any case, the water you finally drink is close to neutral, not because it started out that way, but because it is treated with chemicals to correct the acid or alkali content it originally had.

TIN-CAN TELEGRAPH

TIN-CAN TELEGRAPH

The telegraph was invented in 1837 by Samuel F.B. Morse. This simple gadget first made it possible to receive messages within an instant after they were sent! Take about 30 minutes to build your own telegraph and discover how messages can be relayed by electricity.

YOU WILL NEED

1 #303 Tin can (such as a 1-pound vegetable can)
Can opener
1 10-Foot length of #20 insulated wire
1¼" Steel bolt about 3" long
Pliers
Scissors
1 "D" battery
Sandpaper, fine or medium grade
Masking tape

A telegraph operates on *electromagnetism.* A tiny magnetic field is formed around a wire when electricity flows through it. By wrapping the wire around a steel bolt, the bolt will become magnetic, too— once the electrical circuit has been completed, or *closed.* You can arrange it so that each time the circuit is closed, the magnet attracts a piece of steel and they come together with a loud *click!* And that's what makes a telegraph!

1 Use a can opener to cut the top of the tin can completely off. *Save the top!* Clean the can and remove the label. Cut the bottom *most* of the way off, but not completely: Leave a ¾" section uncut.

2 Use pliers to bend the edges of the still-connected lid, as shown: first the two edges, then the tip. This is your flapper. Then squeeze the can (fingers should be under the flapper and on the side opposite where it is still attached to the can) until the can becomes an oval shape. Bend the flapper down so that it sticks out over the edge of the can.

3 Cut 6" of wire from the 10-foot length. Strip about 1" of the insulation from both ends of the long *and* short pieces of wire. (To do this, use scissors to cut through the plastic insulation, applying a *very gentle* pressure so that you won't cut all the way through to the wire.) Wrap the long wire around the bolt *very neatly,* placing one loop right next to the last one wrapped. Make sure you leave about 6" of each end of the wire sticking out. Wrap some tape around the wound-up wire to keep it in place.

4 Securely tape one end of this long wire to one end of the "D" battery and one end of the short wire to the other end of the battery. Be sure to put plenty of tape on both ends of the battery.

5 Now take the other lid that you removed and saved, and use the pliers to bend back the sharp edges, just as you did for the flapper. *However,* before you do the last edge, pinch the other end of the *short* wire into the fold. **But first:** Use sandpaper to rub away any coating on the can lid, at the spot where the wire will make contact. Now you've made your "telegraph key."

6 Now assemble the telegraph. Tape the bolt to the side of the can so the head of the bolt is right under the tip of the flapper, about ⅛″ away. Wrap the tape around the bolt and can several times. Wedge the battery into the bottom of the can.

7 Tape the key to the side of the can as illustrated. Tape the remaining exposed end of the long wire to the can *under* the free end of the key, but *not* touching it. Now, when you press down on the key, it contacts the bare wire, the circuit is completed, the magnetized bolt pulls down the flapper, and the Tin-Can Telegraph

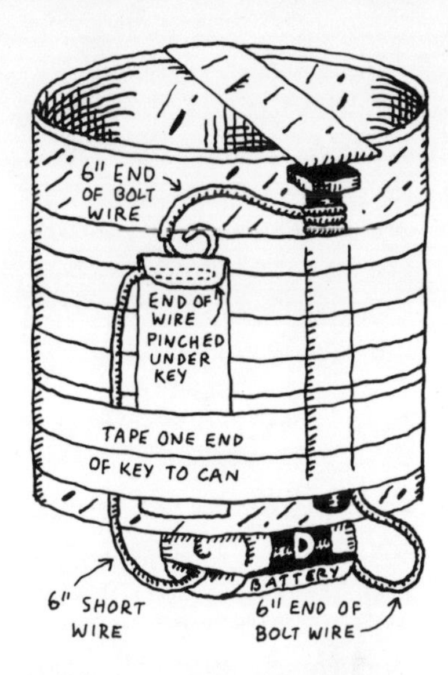

goes *click, click, click!* **Note:** If the flapper sticks to the bolt, increase the distance between the end of the flapper and the head of the bolt. If the flapper won't pull over to the bolt, check your circuit, adjust the gap, or get a new battery.

8 Make up a simple code comprised of clicks (1 click = A, 2 clicks = C, 3 clicks = E, 4 clicks = R, etc.) and take turns using the telegraph to send messages. So, to send the word "race," the sender would key 4 clicks, followed by 1 click, then 2 clicks, and finally 3 clicks.

Actually, old Sam Morse not only popularized the telegraph, he also developed a code for sending messages. The individual clicks sent by the telegraph were separated from one another by either a short interval or a long interval. It was actually the pattern of long and short pauses between clicks that early Morse code users listened to. Today we refer to the code as being made up of *dots* and *dashes.* You can find the Morse code in an encyclopedia or dictionary and use it to send some messages. The easiest way to practice is to use a fingernail on a piece of wood, tapping for the "dots" and scratching for the "dashes."

AFTERWORDS

If you saw some friends waiting down the street, what would you do to get their attention? You would probably yell, maybe whistle, even wave your arms in the air. But what if they were several blocks away? They probably wouldn't be able to hear your shouts or see your arm signals. This is the problem of *telecommunication* (*tele* means "distant"). So how can we communicate at a distance?

One of the earliest forms of telecommunication was pounding on a hollow log with a stick. The next step was to stretch and dry animal skins over the ends of these hollow logs to form drums. A pattern of beats was established to warn of approaching danger or to call people together. But still you had to be close enough to hear the drums to get the message.

Visual signals developed around the same time as sound signals. For example, early American Indians used smoke signals. Messages were also sent by lighting fires on hilltops or beaches. Paul Revere used lighted lanterns to warn of danger during the American Revolution. (Remember "one, if by land; two, if by sea"?) The U.S. Navy still communicates from ship to ship by flashing lights—especially if they don't want to use their radios for fear that the enemy might tune in. The Navy also sends messages between ships with *semaphore* flags: A flag is held in each hand and the different flag positions indicate different letters of the alphabet. Colored lights along railways, near airports, on high buildings and floating buoys, and at traffic intersections are all visual signals that communicate simple messages to us.

But all these light and sound systems of telecommunication are limited by how far you can see or hear. Luckily, once electricity was discovered it wasn't long before electronic ways of telecommunicating were developed. The telegraph, invented by Samuel Morse, was the first practical electronic communicator. Morse was not a skilled scientist, but his curiosity and imagination helped him organize the work of others into a useful machine. In 1843, Congress gave Morse money to run a telegraph line from Washington to Baltimore. Messages "clicked" out with a metal key that could be heard on the receiving end or be printed out onto paper tape. The first telegraphed message was sent on May 24, 1844.

Finally, Alexander Graham Bell developed the telephone in 1876. Both the telegraph and the telephone use electromagnets that need a wire between sender and receiver. But once radio waves (an electromagnetic wave with a radio frequency) were discovered, radio transmitters and receivers were created to send messages through the air *without* wires. Long and complex messages can now be sent all over the world at the speed of light—186,000 miles in 1 second!

Today's modern message magic uses satellites that stay in a fixed position in space. Electromagnetic waves from televised events such as the Olympic Games in Yugoslavia bounce off these satellites and turn up on televisions in American living rooms. We've certainly come a long way from pounding a hollow log!

A SHOOT IN THE DARK

A SHOOT IN THE DARK

A Shoot in the Dark takes only a few minutes to start. It will take about one week to finish the whole experiment.

YOU WILL NEED

1 Bowl or saucer of water
2 Flowerpots or
other containers
12 Bean or pea seeds
(lima, navy, etc.)
Potting soil

It's no accident that plants growing in the sunlight are green. The green color comes from chlorophyll. Together, sunlight and chlorophyll help plants produce their own food in a process called photosynthesis. What happens when a tender, young shoot finds itself trapped in the dark? Try this simple experiment and find out.

1 Place the 12 bean or pea seeds in a bowl or saucer. Soak the seeds in water for a day, until the seed coats wrinkle and crack.

2 Fill the two flowerpots or other containers with soil, leaving about one inch of space at the top of the pot. Place six seeds, evenly spaced, on the surface of the soil. Cover the seeds with about one-half inch of potting soil. Water thoroughly.

3 Place one pot in an undisturbed dark place, such as the back of an unused closet or drawer. It is important to keep this pot in *complete* darkness. Even a few seconds of light can change the results of the experiment.

■ Place the second pot in a sunny window.

4 Let the plants grow for 5 to 10 days. During this time, water the seeds in each pot equally—once every day or two.

■ Remember, you must keep the dark seedlings in total darkness, so be sure the room is completely dark when you open the closet or drawer to water them.

5 When the plants in the sun are about six inches tall and have several leaves, bring the second pot out of the darkness. Place the plants side by side.

6 Compare the plants grown in light with the plants grown in darkness.

■ Are they the same size? If not, which plants are taller?

■ Is there any difference in color?

■ How many leaves does each plant have?

■ How do their roots compare? To find out, carefully dig one plant in each pot out of the soil.

VARIATIONS

■ Put both pots from the experiment into the sunshine. How do the plants grown in the dark change when you put them in a light place?

■ How much light do plants need for normal growth? Grow a few seeds in the dark. Bring them out into the light for 10 minutes a day. How do these plants compare with plants grown completely in the dark or in the light?

■ Try this experiment you can eat. Put several dozen mung beans or alfalfa seeds in a clean jar. Rinse the seeds well, pour off any extra water, and put the jar in a dark place. Rinse the seeds thoroughly twice a day and keep them in the dark until your sprouts are ready. Enjoy them in salads, sandwiches, and Chinese dishes.

AFTERWORDS

It's hard to believe that a big plant can grow from just a tiny seed. It's especially amazing when you see how little of the seed is the new plant. For example, look inside a peanut (or an acorn, or lima bean, or pumpkin or sunflower seed). If you split the peanut in two, you will see the little peanut plant inside. The rest of the peanut is food the plant uses until it's big enough to make its own food through photosynthesis.

When you plant a seed in the ground, it starts to grow very quickly. First, it sends down roots to hold it in place in the soil. Then it sends up a hook-shaped shoot that pushes through the soil easily. A seedling's survival depends entirely upon its ability to find light. Without light to provide the energy for photosynthesis, a plant quickly dies. While the plant is still underground, it has no light. It uses the food inside the seed to grow. As the plant gets bigger, it uses more and more of the food inside the seed. By the time you can see the plant stem coming up through the soil, there isn't much food left in the seed at all.

After the stem comes up through the soil, leaves appear. At an earlier stage, leaves might make it hard for the plant to break through the soil into the sunlight. Once the plant is in sunlight, the green leaves help the plant make its own food through photosynthesis.

In A Shoot in the Dark, the dark closet or drawer acts like the underground conditions of a planted bean. The plants you grew in the dark have the same features as normally grown plants before they push through the soil surface. Their shoot tips are hooked and their shoots are white, elongated, and leafless. Once dark-grown seedlings are moved into the sunlight, leaves and green color quickly develop. Plants need only a short flash of light to begin the developmental processes necessary for photosynthesis.

Botanists (people who study plants) often grow seedlings under dark conditions to study plant growth. Dark-grown seedlings, also called "etoliated seedlings," are also used in the study of plant colorings other than green chlorophyll. More common uses of etoliated seedlings include growing bean sprouts for cooking. Raising etoliated sprouts eliminates the green chlorophyll that gives some plants a bitter taste.

Your findings from this experiment can be applied to your own houseplants. Houseplants often do not receive enough light in homes. Plants that live in poor light become pale, weak, and will not flower. To keep your houseplants healthy, you must provide them with enough sun, or artificial light, or a combination of both. They then will be able to produce enough of their own food to grow and flower. While some plants can grow in the shade, no plant can remain a shoot in the dark for long.

SECRET SIGNALS

SECRET SIGNALS

Make a Secret Signal light to let you know when someone's coming near your room! It will take about one hour to build the light and set it up.

YOU WILL NEED

Standard flashlight that takes "D" batteries
2 "D" batteries
30 Feet of bell wire
Wide adhesive tape, duct tape, or other heavy-duty tape
2 Empty aluminum foil containers from frozen dinners, or 2 aluminum foil pie pans
Several large, thick rubber bands
Scissors, pliers
4 or 5 Pieces of paper, about 3″ × 5″ each

The electricity in your house runs on 120 volts, which is much too dangerous to handle. *Don't ever play around with it.* But you can find out how the electrical circuits in your house work by wiring your own Secret Signal light, which will light up when someone steps on the switch!

1 Decide where you are going to put the switch that will turn on your secret signal light. You could hide it under a rug or mat just outside the door to your room. Once you've decided where to put the switch, measure from that point to the place where you will put the light, and multiply by two. That's how many feet of bell wire you will need. **Note:** Don't try to put the switch more than 15 feet from the light, because the batteries are not strong enough to carry the electricity any farther than that.

2 Cut two pieces of bell wire, each one long enough to stretch from the switch to the light. (But include an extra foot or two of wire to be on the safe side.) Strip about 2″ of plastic insulation away from the ends of both pieces of wire, exposing the bare copper wire inside. To do this, use the scissors to cut through the plastic, applying a *very gentle* pressure so that you won't cut all the way

through the wire. (You might make the first cut as a guide, then rotate the scissors so that the sharp edges cut only the plastic. See illustration **A.**)

3 Take the flashlight apart. You'll see that, at the light bulb end, there's a metal socket holding the bulb. Wrap one piece of wire around the end of the light bulb, making sure that the copper wire touches the metal socket. (See **B.**)

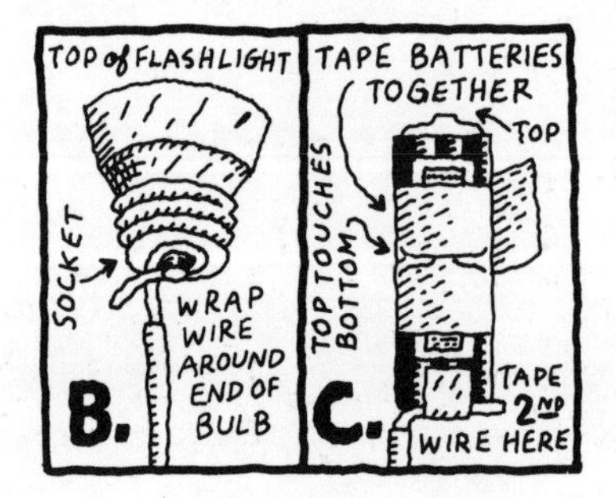

4 Use a wide piece of heavy-duty tape to tape the 2 "D" batteries together as shown in **C**. Wrap the tape *around* the sides of the batteries, but leave the ends of the batteries clear. Be sure the top (positive end) of one battery is touching the bottom (negative end) of the other. Next, tape the second copper wire to the bottom of the battery stack. Then put the batteries together with the top of the flashlight so that the nub on top of the battery touches the end of the light bulb. Tape the batteries to the top of the flashlight by running long pieces of tape down one side, across the bottom, and up the other side of the whole thing. What you've done is to rebuild the flashlight *without* the long part of the outside case.

5 Now you should be able to make the light bulb light up by touching the ends of the two dangling wires together. If it doesn't work, check to be sure your

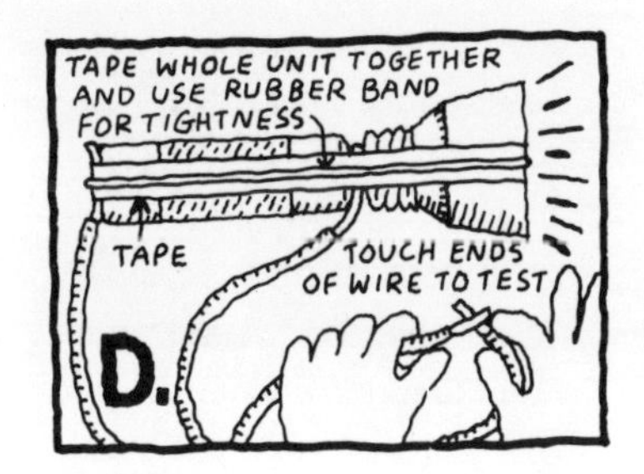

connections are tight. Is the first wire still touching the metal socket around the bulb? Is the second wire pressed tightly against the bottom of the stack of batteries? Wrap one or more long, strong rubber bands around the whole thing to keep the batteries, wire, and light bulb pressed together tightly. (See **D**.)

6 Cut off the sides from two aluminum pie pans or frozen dinner pans, and throw the sides away. With the tip of your scissors, poke a small hole in each pie pan, about ¼″ from the edge. Connect one pie pan to each of the dangling wires attached to the flashlight by putting the copper end of the wire through the hole and bending it around to make a loop. When both pie pans are connected, touch them together and the light will light up. This is your switch.

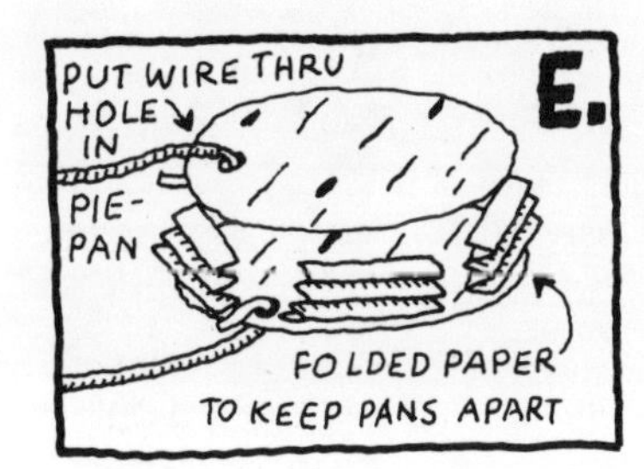

7 Fold over several 3″ × 5″ pieces of paper until you have strips of paper that are several thicknesses thick. Tape the thick paper strips to the top of one pie pan, around the edges. These strips are padding, which will keep the pie pans slightly separated. (See **E**.)

8 Now you're ready to hide the pie pan switch under a rug near your door. Lift the rug and put the padded pie pan down with the paper strips facing up. Put the other pie pan on top of the first one, and replace the rug. The padding should keep the pie pans apart until you step on the rug on top of them. Then their middles will touch (where there's no padding), the circuit will be complete, and the light will light up! Put the light in your room, and tape the wires to the baseboard so they won't show.

VARIATIONS

- Use a dry-cell battery and a small light-bulb socket with two screw terminals on each side of the bulb. That way, you can actually twist the wires around the screws, and the connections won't come loose.
- Be a scientist and invent a way to set up a signal light that will come on when you open a drawer or closet door.
- When you're done using the Secret Signals light, find out how the length of the wire affects the brightness of the bulb. Cut the wires attached to the pie pan switch in half, and strip away the plastic coating. Touch the copper ends together. Is the bulb brighter now? Cut the wires again so that they're only 1 foot long. Why do you think the bulb gets brighter when the wires are shorter?

AFTERWORDS

In Secret Signals you made a *circuit,* a complete circle of wire and/or other substances that would conduct electricity. You saw that if the circuit was broken, the electricity didn't flow. Electricity must *always* flow in an unbroken circle, always returning to its point of origin; that's a primary characteristic of electrical current. To understand why, you have to understand just what makes an electrical current.

The copper atoms in a piece of wire all have the same number of protons, neutrons, and electrons, and this number must always remain the same. It's kind of like the number of seats on a bus. There are only so many seats, and if you have a rule that no one is allowed to stand on the bus, then the bus can only hold a certain number of people. With atoms, you can't add a few extra protons or electrons—the atomic structure has no extra "seats" for them. Besides that, atoms have a rule that says the bus must always be full: If one of the protons, neutrons, or electrons wants to get off the bus, another of the same kind has to get on.

For now, don't worry about the protons and neutrons; they can't move from one atom, or bus, to the next. But the *electrons* in copper can. However, if one electron moves from, let's say, the first bus and gets on the second one, then an electron from the second bus must get off to make room for it—to keep the number of electrons constant. So the electron from the second bus moves to the third one, causing an electron to move from the third bus to the fourth, and so on. This flow of electrons from one atom to another is an electric current.

But this chain reaction can't occur unless there is a circular route, a circuit, for the electrons to take, so that the last atom can pass its "extra" electron back to the first one. Remember, the first atom gave up one of its electrons to start this whole process going. Because the laws of nature say that each atom must always maintain the same number of electrons, the first atom in line must get an electron back. As long as the atoms are arranged in a circular path, the electrons can keep moving from one atom to another.

What if the circular path for electricity was made out of saltwater taffy instead of copper wire? Would the electrons in taffy jump from one to another? No, because taffy isn't a *conductor.* A conductor is any material in which the electrons can move freely from one atom to another. Metals—especially copper and aluminum—are good conductors because their electrons are free to move around. But the electrons in taffy can't, so taffy is called an *insulator.* (But plastic is most commonly used to insulate copper wire.)

This simple example of circuitry extends all the way to the power plant that supplies electricity to your city or town. Even when the power plant is 300 miles from your home, the current must make the trip back to its source to complete the circuit. That's why you always see *two* wires running along the telephone and power poles at the roadside. One wire sends the current out across the countryside. The other wire is the "return" wire, which allows the current to go "home."

NOTES

NOTES

NOTES

NOTES

NOTES